CHEMISTRY for ENVIRONMENTAL and EARTH SCIENCES

CHEMISTRY for ENVIRONMENTAL and EARTH SCIENCES

Catherine V.A. Duke
Craig D. Williams

CRC Press
Taylor & Francis Group
Boca Raton   London   New York

CRC Press is an imprint of the
Taylor & Francis Group, an **informa** business

CRC Press
Taylor & Francis Group
6000 Broken Sound Parkway NW, Suite 300
Boca Raton, FL 33487-2742

© 2008 by Taylor & Francis Group, LLC
CRC Press is an imprint of Taylor & Francis Group, an Informa business

Library of Congress Cataloging-in-Publication Data

Duke, Catherine V. A.
    Chemistry for environmental and Earth sciences / Catherine V.A. Duke and C.D. Williams.
        p. cm.
    Includes index.
    ISBN 978-0-8493-3934-9 (acid-free paper)
    1. Chemistry, physical and theoretical--Environmental aspects--Textbooks. 2. Geochemistry--Textbooks. 3. Environmental chemistry--Textbooks. I. Williams, C. D. (Craig D.) II. Title.

QD453.3.D85 2008
550.1'54--dc22   2007015239

Visit the Taylor & Francis Web site at
http://www.taylorandfrancis.com

and the CRC Press Web site at
http://www.crcpress.com

Table of Contents

Preface

Global warming. Ozone depletion and the hole in the ozone layer. Acid rain. Smog. Water pollution. Contaminated land. These are some of the key issues in environmental science and earth science today. Probably most (if not all) of these terms are already familiar to you from news reports in the media, articles in popular science magazines, scientific programmes on TV, and from the Internet. But in order to be able to understand and tackle these issues, it is necessary to have an understanding of the science behind them, and the science behind all these issues is chemistry. This book, therefore, is a chemistry text book that is aimed specifically at students of environmental science and earth science, and will give them the tools to be able to do just that.

We start with the most fundamental concept of all on which the whole of chemistry is built: atoms and atomic structure. We then develop this by looking at how atoms join together to form molecules, and how those molecules in turn react with each other.

With this understanding of the building blocks of chemistry, we then move on to consider the three "spheres" of the physical world: the geosphere (the solid surface of the Earth), the hydrosphere (the liquid surface of the Earth), and the atmosphere (the gaseous envelope around the Earth), plus the biosphere—the sphere of living organisms. In each case the relevance to environmental science of the chemistry being discussed is highlighted.

Additional material is given in boxes throughout the text. These are intended to develop a few selected topics in more detail.

Throughout the book and at the end of most chapters there are self-assessment questions. These are designed to help you grasp the concepts involved. Although the answers are given in the back of the book, we hope you will try the questions first, before looking at the answers.

The math you will be expected to use in this book will include the basic operations of addition, subtraction, multiplication, and division. In addition, you will be expected to understand and use powers of 10 (exponents) and logarithms. Box 1.1 is a reminder about scientific notation and powers of 10, and Box 1.2 is a reminder about significant figures.

We hope that you enjoy using this book, and that you find it helps you in your study of environmental science or earth science.

Catherine Duke
Craig Williams

Acknowledgments

We would like to thank our colleague Peter Swindells who originally suggested the idea for this book, and started us on our way. Our thanks also to Barbara Hodson for the SEM of the Chalk coccolith.

The Authors

Catherine V. A. Duke obtained her B.Sc. in chemistry from the University of York, U.K., and stayed on at York to do her D.Phil., studying supported inorganic reagents. She did postdoctoral research at Brock University (St. Catharines, Canada) and then worked at Contract Chemicals (Merseyside, U.K.). She joined the University of Wolverhampton as a lecturer in chemistry in 1992.

Craig D. Williams obtained his B.Sc. in chemistry at the University of Salford in the United Kingdom. He then continued at Salford for his M.Sc. and Ph.D., studying the environmental aspects of zeolites. This was followed by postdoctoral research at Edinburgh University on aluminophosphates, and at Liverpool University where he studied zeolite catalysis. He started as a lecturer in chemistry at Wolverhampton in 1990.

1 Fire

In this chapter we will introduce you to the fundamental concepts of chemistry that you will need in your study of earth science or environmental science. We therefore start at the very beginning by considering what atoms are, their structure, how they were formed, and how some atoms are changed by radioactive decay. We then consider how atoms combine to form molecules, the different ways in which atoms bond to one another, and the structures of some inorganic and organic compounds, with particular reference to those of environmental interest, such as pollutants. We then discuss how molecules react with one another in chemical reactions. This chapter also includes a section on the scientific units that you will encounter in chemistry.

1.1 ATOMS AND ELEMENTS

An **element** is the simplest substance that can exist free in nature. Examples of elements include gold particles deposited in streams, carbon in the form of diamonds, and nitrogen and oxygen, which together make up 99% of the Earth's atmosphere. An **atom** is the simplest particle of an element, and atoms thus form the building blocks of all matter as we know it on Earth. Atoms cannot change except by **radioactive decay** (which you will learn about in Section 1.1.5) or in **nuclear reactions**, such as those that take place in the Sun (Section 1.1.2), in nuclear reactors, or in nuclear bombs.

1.1.1 THE STRUCTURE OF ATOMS

An **atom** is made up of three types of particle: **protons, neutrons,** and **electrons**. Protons and neutrons are contained within a **nucleus** at the centre of the atom, whereas electrons orbit the nucleus (Figure 1.1); in fact, most of the volume of an atom is empty space. The nucleus is very small, but nevertheless it contains essentially all the mass of the atom. Within the nucleus are protons, which have a relative mass of 1 and carry a single positive charge, and neutrons, which also have a relative atomic mass of 1 but are electrically neutral (i.e., they have no electrical charge); it is their job to hold the positively charged protons close to one another. The energy that holds the nucleus together is called the **binding energy**, and we will hear more about this later. Circling around the nucleus in specific **orbitals** are electrons. These have a relative atomic mass of only 1/1836 (i.e., 1/1836th the mass of a proton or neutron) and carry a single negative charge. The number of electrons surrounding the nucleus of an atom equals the number of protons in the nucleus. The negatively charged electrons move rapidly through the available atomic volume, held there by

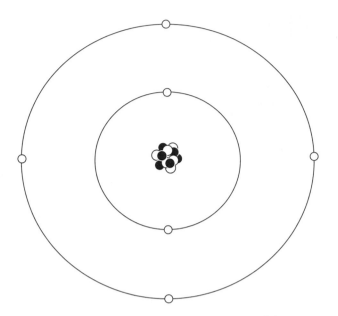

FIGURE 1.1 The structure of the atom: showing the nucleus containing neutrons (larger open circles) and protons (filled circles) with electrons (smaller open circles) in orbitals around the nucleus.

attraction to the positively charged nucleus. Table 1.1 summarises the properties of the three particles found in atoms.

The **atomic number** of an element, which is denoted by the symbol Z, is the number of protons in the nucleus of each of its atoms. All atoms of a particular element have the same atomic number, and each element has a different atomic number from that of any other element, and therefore it is the atomic number that defines an element. For example, the atomic number of sulfur is 16, so it has 16 protons in its nucleus, and 16 electrons orbiting the nucleus. The total number of protons and neutrons in the nucleus of an atom is its **mass number**, which is denoted by the symbol A; each proton and each neutron contributes one unit to the mass number. As the mass number is the sum of protons and neutrons, the number of neutrons, N, equals the mass number minus the atomic number, i.e., $N = A - Z$. For example, the element fluorine has $A = 19$ and $Z = 9$, and therefore $N = 19 - 9 = 10$.

Each element is represented by its **atomic symbol** or **element symbol**, which is a one- or two-letter symbol usually based on the element name (although sometimes the connection between name and symbol is not obvious if the symbol is taken from a non-English name for the element). Table 1.2 is an alphabetical list of all the elements known to date (2007) with their symbols

TABLE 1.1
Properties of Nuclear Particles

Particle	Relative Mass	Charge
Proton	1	+1
Neutron	1	0
Electron	1/1836	−1

TABLE 1.2
The Chemical Elements

Element	Symbol	Atomic Number	Element	Symbol	Atomic Number
Actinium	Ac	89	Mendelevium	Md	101
Aluminium	Al	13	Mercury	Hg	80
Americium	Am	95	Molybdenum	Mo	42
Antimony	Sb	51	Neodymium	Nd	60
Argon	Ar	18	Neon	Ne	10
Arsenic	As	33	Neptumium	Np	93
Astatine	At	85	Nickel	Ni	28
Barium	Ba	56	Niobium	Nb	41
Berkelium	Bk	97	Nitrogen	N	7
Berylium	Be	4	Nobelium	No	102
Bismuth	Bi	83	Osmium	Os	76
Bohrium	Bh	107	Oxygen	O	8
Boron	B	5	Palladium	Pd	46
Bromine	Br	35	Phosphorus	P	15
Cadmium	Cd	48	Platinum	Pt	78
Caesium	Cs	55	Plutonium	Pu	94
Calcium	Ca	20	Polonium	Po	84
Californium	Cf	98	Potassium	K	19
Carbon	C	6	Praseodymium	Pr	59
Cerium	Ce	58	Promethium	Pm	61
Chlorine	Cl	17	Protactinium	Pa	91
Chromium	Cr	24	Radium	Ra	88
Cobalt	Co	27	Radon	Rn	86
Copper	Cu	29	Rhenium	Re	75
Curium	Cm	96	Rhodium	Rh	45
Darmstadtium	Ds	110	Roentgenium	Rg	111
Dubnium	Db	105	Rubidium	Rb	37
Dysprosium	Dy	66	Ruthenium	Ru	44
Einsteinium	Es	99	Rutherfordium	Rf	104
Erbium	Er	68	Samarium	Sm	62
Europium	Eu	63	Scandium	Sc	21
Fermium	Fm	100	Seaborgium	Sg	106
Fluorine	F	9	Selenium	Se	34
Francium	Fr	87	Silicon	Si	14
Gadolinium	Gd	64	Silver	Ag	47
Gallium	Ga	31	Sodium	Na	11
Germanium	Ge	32	Strontium	Sr	38
Gold	Au	79	Sulfur	S	16
Hafnium	Hf	72	Tantalum	Ta	73
Hassium	Hs	108	Technecium	Tc	43
Helium	He	2	Tellurium	Te	52
Holmium	Ho	67	Terbium	Tb	65

Continued

TABLE 1.2 *(Continued)*
The Chemical Elements

Element	Symbol	Atomic Number	Element	Symbol	Atomic Number
Hydrogen	H	1	Thallium	Tl	81
Indium	In	49	Thorium	Th	90
Iodine	I	53	Thulium	Tm	69
Iridium	Ir	77	Tin	Sn	50
Iron	Fe	26	Titanium	Ti	22
Krypton	Kr	36	Tungsten	W	74
Lanthanum	La	57	Uranium	U	92
Lawrencium	Lr	103	Vanadium	V	23
Lead	Pb	82	Xenon	Xe	54
Lithium	Li	3	Ytterbium	Yb	70
Lutetium	Lu	71	Yttrium	Y	39
Magnesium	Mg	12	Zinc	Zn	30
Manganese	Mn	25	Zirconium	Zr	40
Meitnerium	Mt	109			

and atomic numbers. Information about the nuclear mass and charge is often included with the element symbol. The atomic number (Z) is written as a left subscript, and the mass number (A) is a left superscript, so an element with element symbol X would appear as $^{A}_{Z}X$. For example, fluorine (whose symbol is F), which has $A = 19$ and $Z = 9$, is written as $^{19}_{9}F$.

All atoms of a particular element are identical in atomic number, but different atoms of the same element can have different mass numbers. For example, all carbon atoms (symbol C) have 6 protons in the nucleus ($Z = 6$), but only 98.89% of naturally occurring carbon atoms have 6 neutrons in the nucleus ($A = 12$). A small percentage (1.11%) has 7 neutrons in the nucleus ($A = 13$), and a very few (less than 0.01%) have 8 ($A = 14$). These are examples of **isotopes**. Isotopes of an element are atoms that have different numbers of neutrons and, therefore, different mass numbers, so the three isotopes of carbon are written as $^{12}_{6}C$, $^{13}_{6}C$, and $^{14}_{6}C$. All of these carbon isotopes have 6 protons and 6 electrons. In practice, the atomic number is usually not included, because it can be inferred from the element symbol. All isotopes of an element have nearly identical chemical behaviour, even though they have different masses. (However, the very slight differences in chemical behaviour of ^{12}C and ^{13}C and physical properties of ^{16}O and ^{18}O, for example, have enabled geologists to study environmental conditions in recent and ancient Earth history.)

The mass of an atom is measured most easily relative to the mass of a chosen atomic standard. The modern atomic standard is the carbon-12 atom, ^{12}C, and its mass is defined as exactly 12 atomic mass units (amu). Thus, the atomic mass unit, which is also given the name dalton, Da, is 1/12th the mass of the carbon-12 atom. The actual mass of a dalton or amu is 0.0000000000000000000000016605402 g. Such small numbers can appear cumbersome, and so scientists use a method called scientific notation to display very large or very small numbers. In scientific notation the mass of a dalton would be written as 1.66054×10^{-24}g. Read Box 1.1 on p. 20 if you are unfamiliar with scientific notation or if you need a reminder.

The **atomic mass** (also called **atomic weight**) of an element is the average of the masses of its naturally occurring isotopes weighted according to their abundances. For example, the atomic mass of chlorine is 35.45 amu or 35.45 Da, which is the weighted average of its two isotopes, ^{35}Cl (34.97 amu, 75.8% of chlorine atoms) and ^{37}Cl (36.97 amu, 24.2% of chlorine atoms). The atomic mass of chlorine can be calculated as follows:

$$\text{Atomic mass of chlorine} = (34.97 \times 75.8/100) + (36.97 \times 24.2/100)\ \text{amu} = 35.45\ \text{amu}$$

SELF-ASSESSMENT QUESTIONS

Q1.1   Use the periodic table in Figure 1.2 to identify the following elements:
    (i) $^{11}_{5}R$
    (ii) $^{20}_{10}A$
    (iii) $^{23}_{11}T$
    (iv) $^{40}_{20}E$
    (v) $^{75}_{33}G$
    (vi) $^{89}_{39}Q$
    (vii) $^{103}_{45}X$
    (viii)$^{181}_{73}M$
    (ix) $^{209}_{84}Z$

Q1.2   Using the periodic table as a guide, identify the following isotopes:
    (i) $^{117}_{50}T,\ ^{118}_{50}T,\ ^{119}_{50}T$
    (ii) $^{28}_{14}X,\ ^{29}_{14}X,\ ^{30}_{14}X$
    (iii) $^{236}_{92}Z,\ ^{237}_{92}Z,\ ^{238}_{92}Z$

Q1.3   For elements T, X, and Z, calculate the number of neutrons (N) in each isotope:
    (i) $^{117}_{50}T,\ ^{118}_{50}T,\ ^{119}_{50}T$
    (ii) $^{28}_{14}X,\ ^{29}_{14}X,\ ^{30}_{14}X$
    (iii) $^{236}_{92}Z,\ ^{237}_{92}Z,\ ^{238}_{92}Z$

Q1.4   Calculate the atomic mass of copper, Cu, which has two isotopes: ^{63}Cu (62.93 amu, 69.5% of copper atoms) and ^{65}Cu (64.93 amu, 30.5% of copper atoms).

1.1.2 THE ORIGIN OF THE ELEMENTS

How did the universe begin? How were the elements formed? It is only recently that these questions have begun to be answered. The currently accepted model proposes that a sphere of unimaginable properties (diameter 10^{-30} m, density 10^{99} kg m^{-3}, and temperature 10^{32} K) exploded in a **big bang** about 13.7 billion years ago. One second later, the universe was an expanding mixture of protons, neutrons, and electrons, denser than gold and hotter than an exploding hydrogen bomb. During

1	2	3	4	5	6	7	8	9	10	11	12	13	14	15	16	17	18
1 H 1.0079																	2 He 4.0026
3 Li 6.941	4 Be 9.0122											5 B 10.811	6 C 12.011	7 N 14.007	8 O 15.999	9 F 18.998	10 Ne 20.180
11 Na 22.990	12 Mg 24.305											13 Al 26.982	14 Si 28.086	15 P 30.974	16 S 32.065	17 Cl 35.453	18 Ar 39.948
19 K 39.098	20 Ca 40.078	21 Sc 44.956	22 Ti 47.867	23 V 50.942	24 Cr 51.996	25 Mn 54.938	26 Fe 55.845	27 Co 58.933	28 Ni 58.693	29 Cu 63.546	30 Zn 65.409	31 Ga 69.723	32 Ge 72.64	33 As 74.922	34 Se 78.96	35 Br 79.904	36 Kr 83.798
37 Rb 85.468	38 Sr 87.62	39 Y 88.906	40 Zr 91.224	41 Nb 92.906	42 Mo 95.94	43 Tc [97.907]	44 Ru 101.07	45 Rh 102.91	46 Pd 106.42	47 Ag 107.87	48 Cd 112.41	49 In 114.82	50 Sn 118.71	51 Sb 121.76	52 Te 127.60	53 I 126.90	54 Xe 131.29
55 Cs 132.91	56 Ba 137.33	57–71 Lanthanides	72 Hf 178.49	73 Ta 180.95	74 W 183.84	75 Re 102.91	76 Os 190.23	77 Ir 192.22	78 Pt 195.08	79 Au 196.97	80 Hg 200.59	81 Tl 204.38	82 Pb 207.2	83 Bi 208.98	84 Po [208.98]	85 At [209.99]	86 Rn [222.02]
87 Fr [223]	88 Ra [226]	89–103 Actinides	104 Rf [261]	105 Db [262]	106 Sg [266]	107 Bh [264]	108 Hs [277]	109 Mt [268]	110 Ds [271]	111 Rg [272]							

Lanthanides

57 La 138.91	58 Ce 140.12	59 Pr 140.91	60 Nd 144.24	61 Pm [145]	62 Sm 150.36	63 Eu 151.96	64 Gd 157.25	65 Tb 158.93	66 Dy 162.50	67 Ho 164.93	68 Er 167.26	69 Tm 168.93	70 Yb 173.04	71 Lu 174.97

Actinides

89 Ac [227]	90 Th 232.04	91 Pa 231.04	92 U 238.03	93 Np [237]	94 Pu [244]	95 Am [243]	96 Cm [247]	97 Bk [247]	98 Cf [251]	99 Es [252]	100 Fm [257]	101 Md [258]	102 No [259]	103 Lr [262]

FIGURE 1.2 The periodic table of the elements.

the next few minutes, it became a gigantic fusion reactor creating the first atomic nuclei, hydrogen (^{1}H and ^{2}H), and helium (^{3}He and ^{4}He). After 10 minutes, more than 25% of the mass of the universe existed as ^{4}He; most of the remainder was ^{1}H, and about 0.001% was ^{2}H. About 100 million years later (13.6 billion years ago), gravitational forces pulled this cosmic mixture into primitive contracting stars.

Analysis of electromagnetic radiation (typically ultraviolet, visible, or infrared radiation) being emitted by stars, and chemical analysis of the Earth, Moon, and meteors furnishes data about isotope abundance in the universe. From these have been developed a model for **stellar nucleogenesis**, the origin of the elements in the stars. The process involves "burning," but this is very different from the burning we are familiar with. In stellar nucleogenesis, burning means the conversion of atomic nuclei from one element to another, and involves the **fusion** of atomic nuclei or the capture of a nuclear particle (proton, neutron, or electron). These processes are accompanied by the release of energy and, sometimes, the emission of other nuclear particles.

The first stage in the process is **hydrogen burning** (H burning), which produces helium. The initial contraction of a star heats its core to about 10^7 K, at which point H burning starts. In this process, a helium nucleus (^{4}He) is produced from the fusion of four hydrogen nuclei (i.e., protons). Two **positrons** (particles with the same mass as electrons but carrying a single positive charge, and written as e$^+$) and two **neutrinos** (nearly massless particles shown as v) are emitted, and a great amount of energy is released in the process. This is shown in Equation 1.1.

$$4{}_1^1\text{H} \longrightarrow {}_2^4\text{He} + 2\text{e}^+ + 2\text{v} + \text{energy} \qquad (1.1)$$

After several billion years of H burning, about 10% of the ^{1}H is consumed, and the star contracts, starting the second stage in nucleogenesis, which is **helium burning** (He burning). In this process, carbon (C), oxygen (O), neon (Ne), and magnesium (Mg) are formed. The ^{4}He produced in H burning forms a dense core, hot enough (2×10^8 K) to fuse ^{4}He. The energy released during helium burning expands the remaining ^{1}H into a vast envelope, and the star becomes a red giant, more than 100 times its original diameter. Within its core, pairs of ^{4}He nuclei, which are also known as α (alpha) particles, fuse into an unstable ^{8}Be nucleus, which then collides with a third ^{4}He to form stable a ^{12}C nucleus. Subsequent fusion of ^{12}C with more ^{4}He creates nuclei up to ^{24}Mg, as shown in Equation 1.2.

$$^{12}\text{C} \xrightarrow{\ \alpha\ } {}^{16}\text{O} \xrightarrow{\ \alpha\ } {}^{20}\text{Ne} \xrightarrow{\ \alpha\ } {}^{24}\text{Mg} \qquad (1.2)$$

After a few million years, all the ^{4}He is consumed, and heavier nuclei form the core, which contracts and heats, expanding the star to a supergiant. In the hot core (which has a temperature of 7×10^8 K), carbon and oxygen burning occur, as shown in Equations 1.3 and 1.4. Other reactions occur in which very energetic α particles are released. These α particles can then form nuclei up to ^{40}Ca as shown in Equation 1.5.

$$^{12}\text{C} + {}^{12}\text{C} \longrightarrow {}^{23}\text{Na} + {}^{1}\text{H} \tag{1.3}$$

$$^{12}\text{C} + {}^{16}\text{O} \longrightarrow {}^{28}\text{Si} \tag{1.4}$$

$$^{12}\text{C} \xrightarrow{\alpha} {}^{16}\text{O} \xrightarrow{\alpha} {}^{20}\text{Ne} \xrightarrow{\alpha} {}^{24}\text{Mg} \xrightarrow{\alpha}$$
$$^{28}\text{Si} \xrightarrow{\alpha} {}^{32}\text{S} \xrightarrow{\alpha} {}^{36}\text{Ar} \xrightarrow{\alpha} {}^{40}\text{Ca} \tag{1.5}$$

Further contraction and heating of the core to 3×10^9 K allows reactions in which nuclei release neutrons, protons, and alpha particles and then recapture them. As a result, nuclei with lower binding energies supply protons and neutrons to create nuclei with higher binding energies. The process, which takes only minutes, stops at iron (Fe, $A = 56$) and nickel (Ni, $A = 58$), the nuclei with the highest binding energies.

In very massive stars, the next stage is the most spectacular, and this is where the elements heavier than iron and nickel form. With all the fuel consumed, the core collapses within a second. Many iron and nickel nuclei break down into neutrons and protons. Protons capture electrons to form neutrons, and the entire core forms an incredibly dense neutron star (a sun-sized star that became a neutron star would fit in the Greater London area). As the core implodes, the outer layers explode into a supernova, which expels material throughout space. A supernova occurs an average of every few hundred years in each galaxy. The heavier elements form during supernovas and are found in second-generation stars, those that coalesce from inter-stellar ^{1}H and ^{4}He and the debris of exploded first-generation stars.

Heavier elements form through neutron-capture processes. In the slow neutron absorption process (**s-process**), a nucleus captures a neutron (written as $^1_0$n) and emits a γ (gamma) ray. Days, months, or even years later, the nucleus emits an electron, also known as a β (beta) particle, as the neutron converts to a proton. This increases the atomic number by one to form the next element, as in the conversion of $^{68}_{30}$Zn to $^{69}_{31}$Ga shown in Equation 1.6.

$$^{68}_{30}\text{Zn} + {}^{1}_{0}\text{n} \longrightarrow {}^{69}_{31}\text{Ga} + e^{-} \tag{1.6}$$

The stable isotopes of most heavy elements form by the s-process. Less-stable isotopes and those with mass numbers greater than 230 cannot form by the s-process because their **half-lives** are too short. (A half-life is the time taken for half of the atoms to decay to another atom.) These form by the rapid neutron absorption process (**r-process**) during the fury of a supernova. Multiple neutron captures, followed by multiple β decays, occur in a second, as when ^{56}Fe is converted to ^{79}Br by gaining 23 neutrons, as shown in Equation 1.7.

$$^{56}_{26}\text{Fe} + 23^{1}_{0}\text{n} \longrightarrow {}^{79}_{35}\text{Fe} \longrightarrow {}^{79}_{35}\text{Br} + 9e^{-} \tag{1.7}$$

1.1.3 THE PERIODIC TABLE

At the end of the 18th century the French chemist Antoine Lavoisier compiled a list of the 23 elements known at that time. By 1870 there were 65 known elements, by 1925 a total of 88 were known, and today there are 111 (including some very short-lived elements artificially formed in nuclear colliders). As the number of known elements grew, some way of classifying them was urgently required.

By the mid-19th century, enormous amounts of information concerning reactions, properties, and atomic masses of the known elements had accumulated. Several researchers noted recurring or periodic patterns of behaviour and proposed schemes to organize the elements according to some fundamental property. In 1871, the Russian scientist Dmitri Mendeleev published the most successful organising scheme, a table that listed the elements by increasing atomic mass, arranged so that elements with similar chemical properties were put in the same column. One of Mendeleev's great achievements was the prediction of the existence of then-unknown elements (for example, gallium and germanium) by leaving gaps in his table where no known element seemed to fit. Mendeleev's original periodic table was later modified by the English scientist H.G.J. Moseley, who ordered the elements in terms of their atomic number instead of atomic mass. The modern periodic table of the elements (based on Mendeleev's original version as modified by Moseley) is one of the great classification schemes in science.

A modern version of the periodic table is shown in Figure 1.2. The layout of the periodic table is as follows:

1. Each element is in a box that contains its atomic number, atomic symbol, and atomic mass. The boxes lie in order of increasing atomic number moving from left to right.
2. The boxes are arranged into a grid of **periods** (horizontal rows) and **groups** (vertical rows). Each period is numbered from 1 to 7. Each group is numbered from 1 to 18.
3. Groups 1 and 2 form the **s block**. Groups 13–18 form the **p block**. Together groups 1, 2, and 13–18 contain the **main group elements**. Groups 3–12 form the **d block** and contain the **transition elements**. Two horizontal series of **inner transition elements**, the **lanthanides** (or **lanthanoids**) and **actinides** (or **actinoids**), fit between the elements of group 3 and 4 in the 6th and 7th periods, and are usually placed below the main body of the table. These form the **f block**.

At this stage, the clearest distinction among the elements is their classification as **metals**, **nonmetals**, or **metalloids**. The elements highlighted in grey in the periodic table are the metalloids, forming a diagonal line across the p block and separating the metals from nonmetals. Very little astatine (At) exists, as all its isotopes are radioactive with short half-lives (see Section 1.1.5). It possesses some metallic character, but also has some similar properties to iodine (I), a nonmetal. About three-quarters of the elements are metals, and these appear in the large left and lower portion of the table. The nonmetals appear in the small upper-right portion of the table.

Several groups have special names. The elements of group 1 are given the name **alkali metals**, those of group 2 are called the **alkaline earth metals**, group 15 are sometimes referred to as the **pnictogens**, group 16 are called the **chalcogens**, group 17 are the **halogens**, and group 18 are the **noble gases**.

You may see different numbering schemes for the periodic table groups in other (especially older) texts. For example, groups may be numbered 1A-8A (with 8A covering groups 8-10) followed by 1B-8B; or the main groups (groups 1, 2, and 13–18) may be numbered 1A-8A and the transition metals numbered 1B-8B (with 8B covering groups 8–10); or sometimes, only the eight groups of the main group are numbered 1–8 (or even 1–7 followed by 0). The 1–18 numbering scheme used here is that recommended by the International Union of Pure and Applied Chemistry (IUPAC). It has the advantage of being unambiguous by avoiding any confusion over A and B subdivisions.

1.1.4 ELECTRONS AND ELECTRON ORBITALS

It was the Danish physicist Niels Bohr who developed the concept of an atom consisting of a central nucleus around which electrons orbit in different orbitals. One of the observations that helped him to this conclusion was that of **atomic spectral emission**. It was noted that when an electrical current was passed through a tube of hydrogen gas, a series of bright lines was observed in its spectrum (Figure 1.3). When other elements were placed in the tube, they also gave off light lines, but at different wavelengths. The Bohr concept of the atom explains these lines: an electron can jump from one orbital to another and in doing so take in or release energy of specific wavelengths. If the electron jumps closer to the nucleus, it releases energy, which is seen as bright light lines in a spectrum.

The electron orbitals are numbered in sequence from 1 onward moving out from the nucleus, as shown in Figure 1.4. The orbitals themselves consist of various different suborbitals. The simplest suborbital is an **s orbital**, which is spherical in shape. The next simplest suborbitals are the **p orbitals**; there are three p orbitals, all are dumbbell shaped. The next simplest suborbitals are the **d orbitals**. There are five d orbitals, four of which are shaped like crossed dumb-bells, and one that is shaped like a doughnut with a dumb-bell though the centre. The next simplest suborbitals are the **f orbitals**. There are seven f orbitals, and these have rather complicated shapes. The shapes of the s, p, and d orbitals are shown in Figure 1.5.

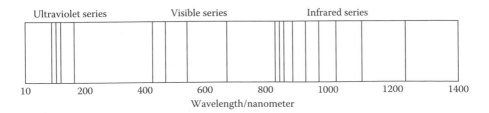

FIGURE 1.3 Spectral lines in atomic emission spectra.

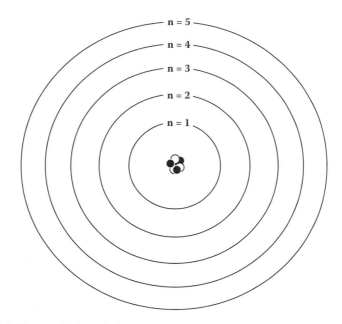

FIGURE 1.4 The numbering of electron orbitals.

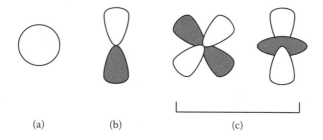

(a)   (b)   (c)

FIGURE 1.5 Electron orbitals: (a) an s orbital, (b) a p orbital, (c) two types of d orbital.

Each s, p, d, or f orbital is capable of holding two electrons, and therefore we can see that each s orbital holds two electrons, the three p orbitals can hold six electrons, the five d orbitals can hold ten electrons, and the seven f orbitals can hold a maximum of fourteen electrons.

The first orbital around a nucleus is called the n = 1 orbital. It consists of an s orbital only, which is labelled as 1s, and it can hold two electrons.

The next stable orbital moving out from the nucleus is larger than the first orbital, and is called the n = 2 orbital. It consists of an s orbital (labelled as 2s) and three p orbitals (labelled as 2p), and can accommodate up to eight electrons (i.e., two in the 2s orbital and six in the 2p orbitals).

The next stable orbital moving out from the nucleus is larger than the second and is called the n = 3 orbital. It consists of an s orbital, three p orbitals, and five d orbitals, labelled as 3s, 3p, and 3d, and can accommodate up to eighteen electrons (i.e., two in the 3s, six in the 3p, and ten in the 3d).

TABLE 1.3
Electron Orbitals

n	Orbitals	Total Number of Electrons
1	1s	2
2	2s, 2p	8
3	3s, 3p, 3d	18
4	4s, 4p, 4d, 4f	32
5	5s, 5p, 5d, 5f	32
6	6s, 6p, 6d, 6f	32

The next stable orbital moving out from the nucleus is larger than the third and is called the n = 4 orbital. It consists of an s orbital, three p orbitals, five d orbitals, and seven f orbitals, labelled as 4s, 4p, 4d, and 4f, and can accommodate up to 32 electrons (i.e., two in the 4s, six in the 4p, ten in the 4d, and fourteen in the 4f).

Moving further out from the nucleus are the n = 5 and n = 6 orbitals. These orbitals consist of the equivalent orbitals to those for n = 4; i.e., there are 5s, 5p, 5d, and 5f orbitals in n = 5, and 6s, 6p, 6d, and 6f orbitals in n = 6, and each of n = 5 and n = 6 can accommodate up to 32 electrons. In theory there are the possibilities of nine **g orbitals** in the n = 5 orbital and eleven **h orbitals** in the n = 6 orbital, but no electrons fill these orbitals. The electron orbitals and the various suborbitals they contain are summarised in Table 1.3.

With the various orbitals established, we need to consider the order in which electrons fill up the various orbitals. Electrons occupy the lowest energy orbitals available as this makes the atoms more stable. In energy terms, the 1s orbital has the lowest energy and is filled first. The sequence of filling is: 1s, 2s, 2p, 3s, 3p, 4s, 3d, 4p, 5s, 4d, 5p, 6s, 4f, 5d, 6p, ..., as shown in Figure 1.6. (For reasons beyond the scope of this book, d and f orbitals are filled after the s and p orbitals from the next orbital.)

The **electron configuration** for each element can be worked out by allocating electrons to orbitals in sequence until all the electrons are used up. We will give the electronic configuration of germanium as an example. Germanium has atomic number 32, and therefore it has 32 electrons. These are allocated as follows:

Two electrons in the 1s
Two electrons in the 2s
Six electrons in the 2p
Two electrons in the 3s
Six electrons in the 3p
Two electrons in the 4s
Ten electrons in the 3d
Two electrons in the 4p
Total = 32 electrons

The numbers of electrons in each of the orbitals is indicated as a superscript and so the electron configuration of germanium is written as:

$$1s^2\ 2s^2\ 2p^6\ 3s^2\ 3p^6\ 4s^2\ 3d^{10}\ 4p^2$$

This configuration tells us, for example, that in the outermost occupied orbital (n = 4) we have $4s^2$ and $4p^2$; i.e., there are four outer, or **valence**, electrons.

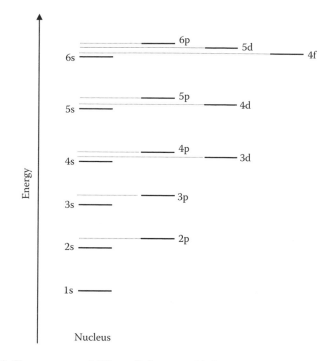

FIGURE 1.6 The sequence of filling of electron orbitals.

The elements of group 18 come at the end of each row of the periodic table, and therefore each of their orbitals is either completely filled or completely empty. This is a particularly stable arrangement of electrons, which is why the group 18 elements are very unreactive and make few chemical compounds. Such an electron configuration is called a **noble gas configuration**. The electronic configurations of the group 18 elements are given in Table 1.4 together with their shorthand forms. The electronic configuration of other elements can be written in a simplified form as the shorthand form of the preceding (by atomic number) group 18 element, followed by the valence electrons of the element under consideration. The electronic configuration of germanium in this form is written as $[Ar]\ 4s^2\ 3d^{10}\ 4p^2$, because

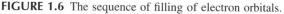

TABLE 1.4
Electronic Configurations of the Group 18 Elements

Element	Configuration	Shorthand
He	$1s^2$	[He]
Ne	$1s^2\ 2s^2\ 2p^6$	[Ne]
Ar	$1s^2\ 2s^2\ 2p^6\ 3s^2\ 3p^6$	[Ar]
Kr	$1s^2\ 2s^2\ 2p^6\ 3s^2\ 3p^6\ 4s^2\ 3d^{10}\ 4p^6$	[Kr]
Xe	$1s^2\ 2s^2\ 2p^6\ 3s^2\ 3p^6\ 4s^2\ 3d^{10}\ 4p^6\ 5s^2\ 4d^{10}\ 5p^6$	[Xe]
Rn	$1s^2\ 2s^2\ 2p^6\ 3s^2\ 3p^6\ 4s^2\ 3d^{10}\ 4p^6\ 5s^2\ 4d^{10}\ 5p^6\ 6s^2\ 4f^{14}\ 5d^{10}\ 6p^6$	[Rn]

argon (atomic number 18) is the last group 18 element before germanium (atomic number 32).

SELF-ASSESSMENT QUESTION

Q1.5   Determine the electronic configurations for the following elements:
   (i)   Be
   (ii)   Si
   (iii) Ca
   (iv)   As
   (v)   Sn

The modern periodic table is classified by atomic number, but because an element has the same number of electrons as it has protons (i.e., as its atomic number), the periodic table is also classified according to the number of electrons each element has and, therefore, according to how those electrons fill the atomic orbitals. To simplify matters, we can ignore filled inner orbitals and, therefore, the periodic table is in effect classified by the number of an element's outer electrons and the orbitals they occupy. The outer electrons are those electrons that take part in chemical reactions, and so the electron configuration of an element determines how reactive that element is. The similarity in chemical properties noted and used by Mendeleev to put elements into columns in his original periodic table reflects an underlying similarity in their arrangement of electrons.

1.1.5 RADIOACTIVITY

In Section 1.1.1, we saw that an atom is composed of a nucleus containing protons and neutrons and orbiting electrons. We also saw that the number of neutrons can be variable, giving rise to isotopes. However, many isotopes are not stable, and some will spontaneously transform from one to another. This process is termed **radioactive decay**, and an isotope that undergoes radioactive decay is termed a radioisotope. The nucleus of a radioisotope is unstable because the total energy content of its nucleus is greater than that of a neighbouring stable nucleus. There are several processes by which a radioisotope can become stable. Radioisotopes of low mass numbers generally attempt to become stable by beta (β) decay. There are two forms of β decay. In beta$^-$ (β^-) decay a neutron (^1_0n) converts into a proton (written as ^1_1p) and an electron (written as e) as shown in Equation 1.8. ^{14}C decays in this manner (Equation 1.9). Because the neutron converts to a proton, the mass number of the isotope does not change, but the atomic number increases by 1.

$$^1_0n \longrightarrow {}^1_1p + e^- \tag{1.8}$$

$$^{14}_6C \longrightarrow {}^{14}_7N + e^- \tag{1.9}$$

TABLE 1.5
Half-Lives of Some Radioisotopes

Isotope	Decay Process	Product	Half-Life
^{238}U	α	^{234}Th	4.5×10^9 years
^{14}C	β^-	^{14}N	5730 years
^{137}Cs	β^-	^{137}Ba	30 years
^{222}Rn	α	^{218}Po	3.8 days
^{15}O	β^+	^{15}N	2.06 min
^{214}Po	α	^{210}Ph	1.64×10^{-4} s

In beta$^+$ (β^+) decay, a proton can decay into a neutron and a positron (a positively charged electron, e$^+$). Again, the mass number does not change, but the atomic number decreases by 1 as there is one less proton in the nucleus (Equation 1.10).

$$\,^1_1 p \longrightarrow \,^1_0 n + e^+ \tag{1.10}$$

A proton can also convert into a neutron by capturing an orbiting electron, emitting X-ray radiation in the process (Equation 1.11). This process is termed *electron capture*.

$$\,^1_1 p + e^- \longrightarrow \,^1_0 n + X\text{-rays} \tag{1.11}$$

Some radioisotopes undergo both β^+ decay and electron capture. For example, $^{18}_9 F$ converts to $^{18}_8 O$ by β^+ decay (97%, Equation 1.12) and electron capture (3%, Equation 1.13).

$$\,^{18}_9 F \longrightarrow \,^{18}_8 O + e^+ \tag{1.12}$$

$$\,^{18}_9 F + e^- \longrightarrow \,^{18}_8 O \tag{1.13}$$

Neutron-rich radioisotopes tend to decay by β^- decay, whereas neutron-poor radioisotopes undergo beta$^+$ decay or electron capture. Many radioisotopes require repeated β decays to become stable, such as radioactive $^{140}_{54} Xe$, which undergoes four successive β^- decays to form stable $^{140}_{58} Ce$ (Equation 1.14).

$$\,^{140}_{54} Xe \longrightarrow \,^{140}_{55} Cs \longrightarrow \,^{140}_{56} Ba \longrightarrow \,^{140}_{57} La \longrightarrow \,^{140}_{58} Ce \tag{1.14}$$

For the heavier radioisotopes the number of beta decays would be enormous, and so they undergo a different kind of radioactive decay process. This is known as **alpha (α) decay** and involves the ejection from the nucleus of an α particle (i.e., $^4_2 He$). Plutonium-239 decays to uranium-235 using this process (Equation 1.15).

The loss of $^{4}_{2}\text{He}$ means that the mass number decreases by 4, whereas the atomic number decreases by 2. The α particle is ejected at speed, and due to its mass is the most ionising but the least penetrating form of radiation.

$$^{239}_{94}\text{Pu} \longrightarrow {}^{235}_{92}\text{U} + {}^{4}_{2}\text{He} \tag{1.15}$$

Another form of radioactive decay is that of spontaneous fission. In this process, a heavy radioisotope will split into two medium-weight nuclei, which are called **fission products**. These are commonly radioactive themselves. An example is the fission of ^{252}Cf, which splits with the emission of two neutrons, as shown in Equation 1.16. The atomic numbers of cadmium (48) and tin (50) add up to the atomic number of californium (98) because there is no change in the total number of protons, and the total mass of the products including the two neutrons adds up to 252, the mass number of ^{252}Cf.

$$^{252}_{98}\text{Cf} \longrightarrow {}^{120}_{48}\text{Cd} + {}^{130}_{50}\text{Sn} + 2{}^{1}_{0}\text{n} \tag{1.16}$$

The last form of radioactive decay we will consider is that of gamma (γ) decay. Here, the radioisotopes are in an **excited** (i.e., high-energy) state. The majority of alpha and beta decays result in excited state **daughter nuclei**, and these lose energy to go to their **ground** (i.e., low-energy) state by emitting gamma (γ) radiation. This gamma radiation is actually very-short-wavelength electromagnetic radiation similar to X-rays. An example is the decay of ^{137}Cs by β⁻ decay to ^{137}Ba, which then emits γ radiation to lose energy (Equation 1.17).

$$^{137}_{55}\text{Cs} \longrightarrow {}^{137}_{56}\text{Ba} \text{ (excited state)} \longrightarrow {}^{137}_{56}\text{Ba} + \gamma \tag{1.17}$$

Radioactive decay is a random process, and it is not possible to predict when any one particular atom will decay. The rate at which a particular radioisotope decays is given by its **half-life**, $t_{1/2}$, which is defined as the time taken for half of the nuclei of a given radioisotope in a large sample to decay. The half-life of a radioisotope is related to its instability, and the more unstable it is, the shorter is its half-life. Some example half-lives are given in Table 1.4, and you can see that these range from fractions of a second to many millions of years. ^{238}U has a half-life about the same length of time as the age of the Earth, so that there is now half the amount of ^{238}U than there was when the Earth formed. The level of radioactivity is measured in Becquerels (Bq), which is 1 radioactive disintegration per second.

SELF-ASSESSMENT QUESTIONS

Q1.6   Give the symbol, atomic number, and mass number of the product, X, of the following radioactive decays:

(i) $^{60}_{27}\text{Co} \rightarrow \text{X} + \text{e}^{-}$

(ii) $^{11}_{6}C \rightarrow X + e^+$

(iii) $^{226}_{88}Ra \rightarrow X + {}^4_2He$

(iv) $^7_4Be + e^- \rightarrow X$

Q1.7   The half-life of ^{137}Cs is 30 years. From an original sample of 80 g of ^{137}Cs, how much ^{137}Cs will be left after:

(i)   30 years?

(ii)   60 years?

(iii)   90 years?

Several important nuclei decay in a sequence of α decays called a **radioactive decay series**. Three such nuclei are thorium-232, uranium-235, and uranium-238, and the ultimate stable products of these decays are isotopes of lead (^{208}Pb, ^{207}Pb, and ^{206}Pb, respectively). The radioactive decay series of ^{238}U is shown in Figure 1.7. It is plotted as a graph of mass number against atomic number. As you can see, most of the decays are α decays, but there are also some β⁻ decays. Elements may occur more than once in a particular decay series; for example, Pb, which occurs as unstable,

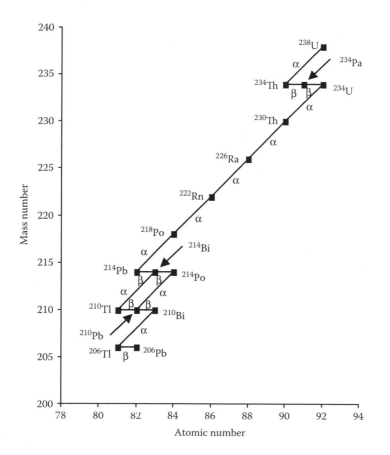

FIGURE 1.7 The radioactive decay series of uranium-238.

radioactive ^{214}Pb and ^{210}Pb before the series ends at stable ^{206}Pb. All the Pb isotopes line up vertically as (by definition) they have the same atomic number.

1.1.6 RADIOMETRIC DATING METHODS

The known, regular rate at which radioisotopes decay is used to date rocks, minerals, and biological objects such as pollen spores and bones. A number of different isotopes can be used for dating, depending on the nature of the material to be dated, e.g., mineral or organic artefact. The age range of material that can be dated is limited by the half-life of the isotope being used because after several half-lives the activity of the radioisotope becomes too low to measure against background radiation. For example, ^{14}C dating cannot be used for objects older than about 50,000 years, which is just under nine half-lives ($t_{1/2}$ for ^{14}C is 5730 years).

Rocks and minerals can be dated using a number of isotopes. The most widely used isotope for dating is ^{87}Rb, which decays to ^{87}Sr by β^- decay (Equation 1.18), because most rocks and minerals contain at least trace quantities of rubidium and strontium. The half-life for this decay is 5×10^{10} years, which is 10 times greater than the age of the Earth. The level of **radiogenic** (formed by radioactivity) ^{87}Sr increases with time, whereas the level of nonradiogenic ^{86}Sr does not change over time. Dating is done by plotting ^{87}Rb/^{86}Sr against ^{87}Sr/^{86}Sr in an **Isochron diagram**.

$$^{87}_{37}\text{Rb} \longrightarrow \, ^{87}_{38}\text{Sr} + e^- \qquad (1.18)$$

Many minerals contain small traces of uranium and thorium, and the decays of ^{232}Th, ^{235}U, and ^{238}U can all be used for dating. These isotopes have complex decay series, which ultimately end with the formation of lead (Equations 1.19–1.21). As these isotopes decay, the amount of radiogenic lead (i.e., ^{206}Pb, ^{207}Pb, and ^{208}Pb) increases relative to nonradiogenic lead (^{204}Pb). The concentration of ^{204}Pb has been constant throughout geological time, and is used as a correction factor. U/Pb dating is achieved by plotting ^{206}Pb/^{238}U against ^{207}Pb/^{235}U in a **Concordia plot**. An example of the application of U/Pb dating is the dating of zircon crystals, which have the chemical formula $ZrSiO_4$. Some zircons are older than 4000 million years, i.e., they are over 88% of the age of the Earth.

$$^{232}\text{Th} \longrightarrow \, ^{208}\text{Pb} \qquad (1.19)$$

$$^{235}\text{U} \longrightarrow \, ^{207}\text{Pb} \qquad (1.20)$$

$$^{238}\text{U} \longrightarrow \, ^{206}\text{Pb} \qquad (1.21)$$

Another isotope important in geological dating is ^{40}K. Potassium occurs in glauconite, which forms on the seabed. Glauconite occurs in a great variety of fossiliferous strata of known geological age from the Cambrian to the present day, and therefore these strata can be dated. For ^{40}K the decay is more complex than ^{87}Rb, and the side effect of this is that the dating is not as accurate as that using ^{87}Rb. About 89% of ^{40}K decays to ^{40}Ca (Equation 1.22), but because ^{40}Ca is the normal isotope of calcium, any minuscule addition from radioactive decay is impossible to distinguish. The remaining 11% of ^{40}K atoms decay to ^{40}Ar by electron capture with a half-life of 11,850 million years (Equation 1.23), and the ratio of ^{40}K to ^{39}Ar is used to date the sample. An improvement on the use of ^{40}K to ^{40}Ar ratios is Ar/Ar dating. In this technique, a sample is irradiated to convert ^{40}K into ^{40}Ar, and the ratio of ^{40}Ar to ^{39}Ar is used to date the sample.

$$^{40}K \longrightarrow {}^{40}Ca + e^- \tag{1.22}$$

$$^{40}K + e^- \longrightarrow {}^{40}Ar \tag{1.23}$$

To construct a time scale based on radioactive decay, it is necessary to have radiometric measurements on rocks or minerals of known geological age across the range of geological time. The rocks that have been used to calibrate the various radioactive dating clocks are:

1. Intrusive igneous rocks and minerals that intersect sedimentary rocks of established stratigraphic position.
2. Accurately dated fossiliferous sedimentary rock that contains **authigenic** (generated in situ) minerals. The radiometric date here is the minimum age of the sedimentary rock, and glauconite is commonly used.
3. Volcanic rocks interbedded with sedimentary rock, whose age is known from the fossil record.
4. Metamorphic rocks, where the minerals date from the time of metamorphism.

By using radiometric dating techniques, it has been possible to determine the age of the Earth. This will be described in Section 2.1. Radiometric dating can also be used to calculate how and when the Earth's atmosphere formed (Section 4.2). Here, we are interested in two radioisotopes: the decay of ^{40}K to ^{40}Ar, and the decay of ^{129}I to ^{129}Xe. ^{40}K has a half-life of 11,850 million years and decays very much more slowly than ^{129}I, which has a half-life of 17.2 million years. This short half-life means that all ^{129}I has decayed, and therefore the amount of ^{129}Xe in the atmosphere is now constant, while the amount of ^{40}Ar continues to increase. Therefore, if the ratio of ^{40}Ar to ^{129}Xe is measured, we can calculate when the atmosphere formed, and also whether it was **outgassed** (expelled from the Earth's mantle) in one go or in stages. The calculations show that 80–85% of our atmosphere was outgassed very early in Earth's history, whereas the last 15–20% has outgassed more slowly over the last 4.4 billion years, mainly through volcanic activity.

The final radiometric dating technique we will consider is that of radiocarbon dating. This was discovered in 1951 when W.F. Libby discovered minute amounts of ^{14}C in air, natural waters, and living organisms. Carbon has three isotopes, which are ^{12}C (98.89%), ^{13}C (1.11%), and ^{14}C. Only ^{14}C is radioactive, and is continuously produced in the atmosphere by cosmic rays. ^{14}C decays by emitting a beta particle and decaying to ^{14}N. The half-life of this process is 5570 years, and it is believed that the production of ^{14}C in the atmosphere has been steady for many thousands of years. The newly formed ^{14}C is oxidised to CO_2 and distributed by wind, rain, rivers, and oceans, giving a constant ^{14}C to ^{12}C ratio in the environment. All living organisms continually absorb CO_2, and so the ^{14}C to ^{12}C ratio stays constant while they are alive. However, once an organism dies the amount of ^{14}C is not replenished, and so begins to decay. The change in the ^{14}C to ^{12}C ratio can then be used to determine when that organism died. Radiocarbon dating is widely used to date objects and events during the latter part of the Ice Age.

Box 1.1 Scientific Notation

Scientific notation is a means of handling very large or very small numbers. It involves expressing a number in the form 1.234×10^x, where x is the power of 10, or **exponent**. In scientific notation, the decimal point always comes after the first digit however many digits there are, so that, for example, neither 123.45×10^2 nor 0.0123×10^3 is correct scientific notation.

For numbers of 10 or higher, x is a positive number and is the number of places the decimal point has to be moved to the *right*. If the decimal point has to be moved more places to the right than there are digits, then zeros are added as required. For example, in the number 1.23×10^4, the decimal place has to be moved four places to the right, but there are only two digits to its right. Therefore two zeros are added after the 3, and the number is 12,300. In the number 1.2345×10^2, the decimal place has to be moved two places to the right, but as there are four digits to the right of the decimal point, no zeros have to be added and the number is 123.45. To convert a number that is larger than 10 to scientific notation, count the number of places that the decimal point has to be moved to the *left* to get it in the correct form. So, for example, the number 456,700 in scientific notation is 4.567×10^5 as the decimal point has to be moved five places to the left.

For numbers less than 1, x is a negative number and is the number of places the decimal point has to be moved to the *left*. The number of zeros that have to be added in front of the first digit is $x - 1$. For example, in the number 1.23×10^{-4}, the decimal point has to be moved four places to the left, and three zeros have to be added in front of the 1, so the number is 0.000123. To convert a number that is lower than 1 to scientific notation, count the number of places the decimal point has to be moved to the *right* to get it in the correct form. So, for example, the number 0.000004567 in scientific notation is 4.657×10^{-6} as the decimal point has to be moved six places to the right.

Sometimes, you will see numbers given simply as powers of 10, for example, 10^4 or 10^{-3}. You need to understand that these are shorthand for 1×10^4 and 1

$\times 10^{-3}$, respectively. This is especially important when using a calculator, as one of the commonest mistakes when inputting a number such as 10^5 is to type in 10EXP5 instead of 1EXP5 (or 10EE5 instead of 1EE5, depending on your calculator), because 10EXP5 is actually 10×10^5, and you will end up with an answer that is 10 times too big.

You will be seeing exponents used throughout this book, and will be expected to perform calculations involving them.

1.2   STATES OF MATTER

In Section 1.1 you learnt that atomic nuclei are made of protons and neutrons, and how the atomic nuclei of different elements are built through the processes of nucleosynthesis in stars and supernovae. The material that results from these processes goes on to form all the matter that we and our environment are composed of. This matter can exist in four forms: solid, liquid, gas, and plasma.

1.2.1   PLASMA

Most matter in the universe is in the unfamiliar form of plasma. In plasma, the atomic nuclei are wholly or largely stripped of their electrons. In this state, matter consists of naked nuclei moving through a sea of electrons (Figure 1.8). Each nucleus can move independently, but may be involved in collisions with other nuclei. This is the state of matter in stars, where the temperature is so high that electrons cannot stay attached to a nucleus because they are constantly knocked off again by

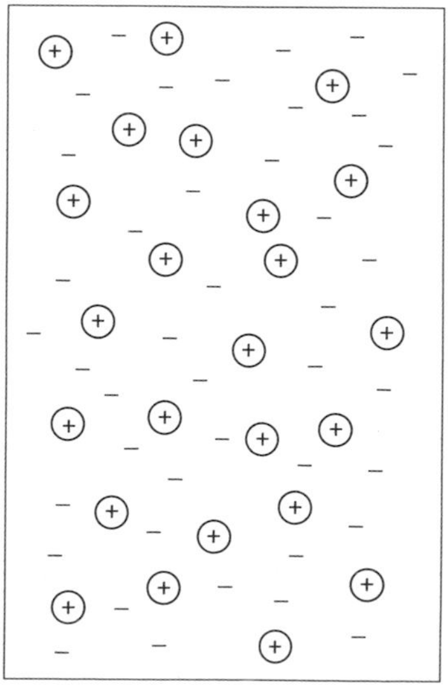

FIGURE 1.8 A plasma.

high-speed collisions with other nuclei and electrons. It is also the state of matter
in much of interstellar space, where radiation from surrounding stars is sufficiently
energetic to knock electrons out of atoms. For example, ionised hydrogen is a major
component of the interstellar medium in our own galaxy, and has also been observed
in other galaxies. In the region of space around the Sun (called the Local Interstellar
Cloud), the density of neutral hydrogen atoms is about 240 atoms per litre, and the
density of H^+ ions is about a third of this, at about 80 atoms per litre. The temperature
in this region is about 7000 K.

In cooler environments (below about 4000 K) and when protected from energetic
radiation, nuclei are largely combined with electrons forming stable atoms. These
conditions exist inside interstellar dust clouds where stars and planets form. Only
in these comparatively rare (for the galaxy) conditions can matter assume the forms
we are most familiar with: gases, liquids, and solids. It is also only under such
conditions that atoms can join together to form **molecules**. (We will consider molecules in more detail in Section 1.4.)

On Earth, plasmas are rare, perhaps the most familiar occurrence of them being
in fluorescent lights. They also occur in lightning bolts, where temperatures are
extremely high and in the upper atmosphere, where the molecules of air are exposed
to high-energy radiation from the Sun. Plasmas in the upper atmosphere give rise
to phenomena such as auroras (see Box 4.1).

1.2.2 GASES

When plasma is cooled and protected from high-energy radiation, electrons combine with the nuclei to form atoms and the atoms may combine together to form
molecules. The result is a **gas**. Gases are like plasmas in that the atoms or
molecules can move independently but unlike them in that they are not electrically
charged. Each gas atom or molecule moves freely, interacting with the other
particles present only when colliding with them or when very close together. The
continual random motion of atoms and molecules in a gas (and also a liquid—
see the next section) is called **Brownian motion**. Of course, unless the gas is
coloured we cannot actually see it moving around, but we can see the effect of
Brownian motion on other particles, for example, in the way that smoke particles
move in air. Because of Brownian motion, and because there are no bonds to hold
the atoms or molecules together in a gas, gases will expand to fill whatever
container they may be placed in, taking up both its shape and its volume (Figure
1.9). We live surrounded by the gases of the Earth's **atmosphere**, which you will
learn about in Chapter 4.

The volume of space, v, occupied by a gas is related to the pressure, p, and
temperature, T, of the gas by the **ideal gas law** (Equation 1.24). In this equation, R
is a constant called the gas constant and n is the number of **moles** of gas. (A mole
is a measure of the amount of a substance, and is defined as 6.022×10^{23} particles;
see Section 1.4.)

$$pV = nRT \qquad (1.12)$$

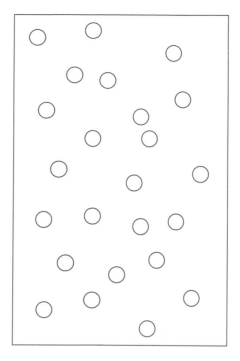

FIGURE 1.9 A gas.

SELF-ASSESSMENT QUESTION

Q1.8 Using the ideal gas law given in Equation 1.24, will the volume of a sample of gas increase or decrease if its temperature increases? (Assume that the pressure is unchanged, and that we are considering a fixed amount of gas, i.e. n is also unchanged.)

The answer to the preceding question (Q1.8) is important with regard to atmospheric meteorological processes. Air that is warmed over land expands and becomes less dense, causing it to rise, whereas air that is cooled in the upper atmosphere contracts, becoming more dense and therefore sinks.

One mole of a gas occupies 22.4 L at atmospheric pressure and 0°C, and therefore 1 litre of a gas at atmospheric pressure contains $6.022 \times 10^{23}/22.4$ particles, which is 2.69×10^{22} particles. Because 1 L of the interstellar medium around the Sun contains only 240 hydrogen atoms, you can see the number of particles in a given volume of gas at atmospheric pressure is very much greater than the number of atomic nuclei in the virtual vacuum of the interstellar medium.

1.2.3 LIQUIDS

Liquids are similar to gases in that the atoms or molecules are able to move relative to one another, and exhibit Brownian motion. Unlike molecules in gases, however,

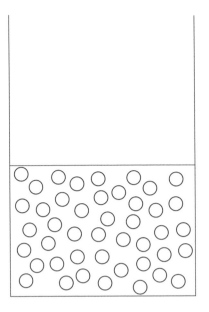

FIGURE 1.10 A liquid.

the molecules of a liquid are in close contact with one another. In a liquid, atoms or molecules are constantly changing their neighbours, sliding past one another to make contact with new atoms or molecules. This gives liquids the ability to flow and to take up the shape of whatever container they are placed in, as shown in Figure 1.10. The commonest liquid on the Earth is water. It is also found in frozen form as ice in glaciers, polar caps, etc. The various forms of water on and beneath the Earth's surface constitute what is called the **hydrosphere**. Water is considered at greater length in Chapter 3.

1.2.4 SOLIDS

A solid consists of an array of atoms, ions (which are charged particles), or molecules in which each atom ion or molecule is in contact with a number of neighbours. Unless they are put under pressure or are held under tension (being pulled apart), the atoms, ions, or molecules in a solid are not free to move relative to one another; if no force is applied, they maintain their shape (Figure 1.11). The atoms, ions, or modules in a solid are often arranged in a highly ordered three-dimensional geometrical pattern. Such an arrangement is called a **crystal**, and the material is said to be **crystalline**. You will be familiar with crystals of such materials as sugar and salt, but many other materials that do not look crystalline are in fact composed of crystals. This is true of many powdered materials, in which the crystals are too small to be seen, and also of metals and many rocks, in which crystals have grown together. In these cases, special techniques may be required to reveal the crystalline nature of the material. You will learn more about the crystalline nature of some of the minerals that make up the Earth in Chapter 2, and you will learn about the crystal structure of some simple inorganic compounds in Section 1.5.

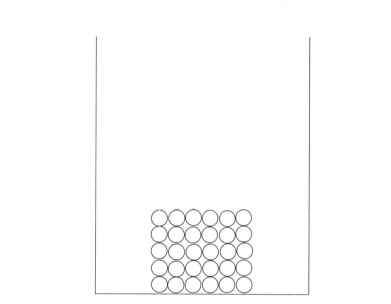

FIGURE 1.11 A solid.

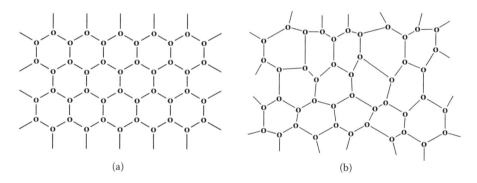

           (a)   (b)

FIGURE 1.12 Crystalline and amorphous solids: (a) A crystalline solid, showing a regular arrangement of atoms, ions, or molecules, in which bond lengths and bond angles are all the same; (b) an amorphous solid showing variable bond lengths and bond angles.

    In some solids, the atoms, ions, or molecules are not in ordered arrays but are arranged randomly. A material that is not crystalline is said to be **amorphous**. Figure 1.12a shows a crystalline arrangement of atoms (or ions or molecules) in a structure, whereas Figure 1.12b shows an amorphous arrangement. In the crystalline material, the distances between the atoms (or ions or molecules) are the same, as are the angles between the bonds joining the atoms. By contrast, in the amorphous material, distances between atoms (or ions or molecules) vary, as do the angles between the bonds. Amorphous materials are often formed by the rapid cooling of a liquid. Usually, when materials freeze (change from a liquid to a solid—see the next section) the atoms, ions, or molecules have time to arrange themselves into the ordered array of a crystal, but sometimes the cooling takes place so quickly that there is insufficient time for this to occur before they are fixed in position, and they can no longer move with respect to one another. In this case, the atoms, ions, or molecules retain the same random arrangement as in the liquid. Such materials are called **glasses**, and an example of this is the mineral obsidian, which is formed as a result of rapid cooling of magma in a volcanic eruption. Glasses are often described as **supercooled liquids** because the random arrangement of the atoms or molecules in the liquid phase has in effect been preserved in the solid phase.

1.2.5 PHASE TRANSITIONS AND PHASE DIAGRAMS

Matter in one phase converts to another phase by a phase transition. This is usually brought about by heating or cooling the material, but can also be caused by a change in pressure. Above a few thousand degrees all matter is in the form of plasma. On cooling the plasma, atoms, and perhaps molecules, form, producing a gas. On further cooling, the gas condenses first to a liquid, and then the liquid freezes to a solid. The phase changes can be observed in reverse order by heating the solid. The temperatures at which these transitions occur are characteristic of the particular material involved, and are frequently used as a means of identifying the material. For example, the melting transition occurs when the atoms, ions, or molecules of the solid acquire sufficient energy to overcome the strength of the attraction between them that make them retain their shape. In a crystalline solid, where the atoms, ions, or molecules are in an ordered array, the required energy will be a certain definite amount and melting will take place at a certain definite temperature. In a glassy solid, where there is no regularity, some bonds will break before others and melting will take place over a range of temperatures. The same is true of mixtures of solids.

The evaporation transition takes place at any temperature. Imagine a sealed evacuated flask at a fixed temperature that contains a quantity of liquid. Some molecules of the liquid close to the surface happen, quite by chance, to get hit from below with sufficient vigour to knock them out of the liquid into the space above, forming a vapour. In this way the number of molecules in the space above the liquid grows. As the number of molecules above the liquid grows, the number of them colliding with the liquid surface and re-entering the liquid phase also grows. Eventually, the numbers entering and leaving the liquid phase in a given time reach a balance, with as many entering as leaving. The concentration of vapour is then constant. The pressure exerted by the vapour at this point is called the **saturated vapour pressure** of the liquid at that temperature. Figure 1.13 shows a sealed container with the same number of molecules returning to the liquid phase as there are going into the vapour phase. Increasing the temperature increases the number of molecules leaving the liquid: the number in the vapour phase will increase until a new equilibrium is reached with an equal number of molecules re-entering the liquid phase. At this new equilibrium, the pressure of the vapour will be greater: the higher the temperature, the higher the saturated vapour pressure. At a sufficiently high temperature, all the liquid will turn into gas.

For a liquid in an open container, molecules of the liquid that pass into the vapour phase due to the vapour pressure of the liquid can get carried away and be permanently removed, and therefore over time the liquid will evaporate away. This

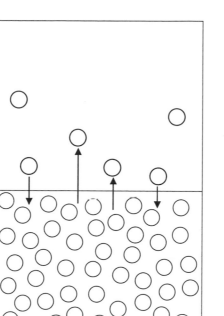

FIGURE 1.13 Saturated vapour pressure of a liquid in a sealed container.

will happen slowly if the liquid has a low vapour pressure or quickly if it has a high vapour pressure (i.e., if it is **volatile**). Boiling occurs in open vessels when the vapour pressure of the liquid becomes equal to the pressure of the surrounding atmosphere. The escaping vapour is then able to push the atmosphere back and form bubbles.

The temperatures at which phase transitions take place in a particular material depend on pressure, and the relationship between pressure, temperature, and the stability of different phases is shown in a **phase diagram**. The phase diagram of carbon dioxide is shown as an example in Figure 1.14. The three areas or fields show the conditions of pressure and temperature at which each phase is the stable phase. The solid lines represent the boundaries between different phases, and mark the conditions under which the phase change from solid and liquid, or liquid and gas, or solid and gas occur. There is one point (i.e., one condition of temperature and pressure) at which solid, liquid, and gas all coexist and are in equilibrium. This is known as the triple point of the liquid, and for CO_2 it occurs at 5.11 atm and $-57°C$. You will note that liquid CO_2 does not exist at atmospheric pressure: when solid CO_2 at atmospheric pressure heats up, it changes directly from the solid state to the gaseous state, without first becoming a liquid. This is known as **sublimation**. For CO_2 to be liquid it has to be at a pressure of at least 5.11 atm, and to be a liquid at room temperature (about 25°C or 298 K), it has to be at a pressure of at least 67 atm. CO_2 cylinders therefore generally contain liquified CO_2 at high pressure. You will consider the phase diagram of water in Chapter 3, and phase diagrams of various silicate minerals in Chapter 2.

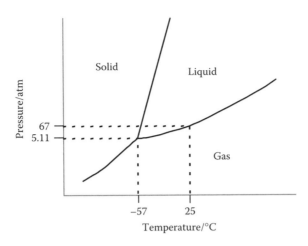

FIGURE 1.14 The phase diagram of CO_2.

SELF-ASSESSMENT QUESTION

Q1.9   What is the stable state of CO_2 at the following conditions of pressure and temperature?
(i)   5.11 atm and –100°C
(ii)   5.11 atm and 25°C
(iii) 67 atm and 0°C

1.2.6 PURE SUBSTANCES, COMPOUNDS, AND MIXTURES

Water, H_2O, is an example of a **pure substance**. It cannot be broken down into simpler components by simple physical separation (e.g., by evaporation, filtration, or dissolution). Similarly, sodium chloride (NaCl, the mineral halite, which is common table salt) is a pure substance as it too cannot be broken down into simpler components by physical means. By contrast, sodium chloride dissolved in water is an example of a **mixture** because the water and sodium chloride can be separated by simple physical means: if the water is evaporated off, the sodium chloride is left behind as a white deposit. Sodium chloride dissolved in water is an example of a **homogenous** mixture: the sodium chloride is dispersed evenly throughout the water, and it is not possible to distinguish visually between the two components.

Another example of a homogeneous mixture is the air that we breathe. It is a mixture of gases that can be separated into its constituent components such as nitrogen, oxygen, and argon, etc., by cycles of compression, cooling, and expansion.

Water is a **compound**: it is formed from the reaction of its constituent **elements**, the gases hydrogen and oxygen, to give a molecule that contains one atom of oxygen and two atoms of hydrogen, as shown in Figure 1.15. The hydrogen and oxygen atoms in water are joined by a **chemical bond**. The resulting compound has very different properties from the elements from which it is formed, so, for example, whereas water is a colourless nonflammable liquid, hydrogen is a colourless highly

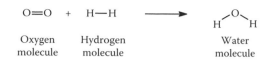

Oxygen   Hydrogen   Water
molecule   molecule   molecule

FIGURE 1.15 The formation of water.

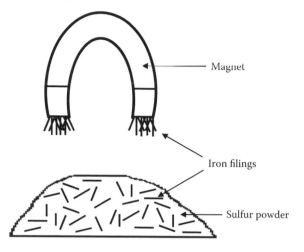

Magnet

Iron filings

Sulfur powder

FIGURE 1.16 Separation of iron filings from sulfur powder.

flammable gas and oxygen is a colourless highly reactive gas. Although water cannot be broken down into its components by physical means, it can be broken down into its constituent elements by chemical means, e.g., by passing an electric current through it or reacting it with particular compounds. Sodium chloride is also a compound: its constituent elements are chlorine (a greenish, poisonous, corrosive gas) and sodium (a soft, reactive metal), which have very different properties from the white, water-soluble, crystalline solid that they form. You will learn about the bonds that form chemical compounds in Section 1.4.

Iron filings (short, needlelike lengths of metallic iron) mixed with flowers of sulfur (a fine yellow powder of elemental sulfur) is an example of a **heterogeneous** mixture because the different components (yellow powder and metallic splinters) can be readily distinguished visually. The mixture can also be easily separated, e.g., by holding a magnet over it, as shown in Figure 1.16. This attracts the iron filings to it, leaving the sulfur powder behind. If, however, a mixture of iron filings and sulfur is heated, a chemical reaction occurs. A new compound, iron sulfide, FeS, is formed, and its constituent elements, iron and sulfur, can no longer be separated by simple physical means.

1.3 UNITS OF MEASUREMENT

In this section we will cover some of the essential tools you will need in understanding scientific measurement and in performing chemical calculations. One aspect of measurement and calculation that you are probably already familiar with is that of **significant figures**. Read Box 1.2 on p. 35 if you need a reminder or if you are unsure how these are worked out.

1.3.1 SI AND NON-SI UNITS

The maximum admissible level of nitrate, NO_3^-, in drinking water is 50 mg L^{-1} or 50 ppm (parts per million). Water boils at a temperature of 100°C or 393 K at a pressure of 1 atm or 101.325 kPa. UV radiation that has a wavelength of 290 nm has an energy of 399 kJ mol^{-1}. All these are facts mentioned in this book, but you need to understand what is meant by mg L^{-1}, ppm, °C, K, atm, kPa, nm, or kJ mol^{-1} in order to be able to make sense of these facts. We need to define our scales of measurement so that data can be shared accurately. It is no use, for example, setting the maximum admissible levels of various pollutants in soil, water, or air if we cannot agree what the numbers actually mean.

The International System of units, otherwise known as the Système International (SI), is based on the metric system of measurement started in France in 1790. There are seven fundamental units covering seven fundamental physical quantities, and from these all other metric units can be derived. The seven fundamental units and some derived units are given in Table 1.6. We will encounter the first five fundamental units, and also a number of derived units. In addition we will encounter a few units that can be changed into SI units by a simple conversion factor. Although most scientific bodies recommend the use of SI units, some non-SI units are still very widely used and accepted. In this book we will generally follow the SI system, but will occasionally use non-SI units. You also need to be familiar with some of the non-SI units.

TABLE 1.6
Fundamental and Derived SI Units

Physical Quantity	Name of Unit	Symbol	Definition
Length	metre	m	
Mass	kilogram	kg	
Time	second	s	
Temperature	kelvin	K	
Amount of substance	mole	mol	
Electric current	ampere	A	
Luminous intensity	candela	cd	
Area	square metre	m^2	
Volume	cubic metre	m^3	
Density	kilogram per cubic metre	kg m^{-3}	
Force	newton	N	kg m s^{-2}
Pressure	pascal	Pa	N m^{-2}
Energy	joule	J	N m
Electric potential difference	volt	V	J A^{-1} s^{-1}
Atomic mass	atomic mass unit	amu	1.66054×10^{-27} KG

The SI unit of length is the metre, m. If the distance being measured is much smaller than 1 m, it may be more convenient to use the related units of centimetre, cm, millimetre, mm, etc., as appropriate, and if the distance being measured is much greater than 1 m, it may be more appropriate to use units of kilometre, km. We will see shortly how cm, mm, and km relate to m.

The SI unit of mass is the gram, g, but the related units of milligram, mg, or kilogram, kg, may be more appropriate in particular situations. We will see shortly how mg and kg relate to g.

The SI unit of time is the second, s, but other units may be used such as minute, hour, day, or year as appropriate, for example for the half-lives of radioisotopes. For long geological timescales you may see the units Ma or Ga used, where a is the symbol for year (often used in preference to y), and Ma and Ga are 10^6 years and 10^9 years, respectively.

The SI unit of temperature is the kelvin, K. On the kelvin scale, absolute zero has a temperature of 0 K, ice melts and water freezes at 273.15 K (at standard atmospheric pressure), and water boils at 373.15 K (also at standard atmospheric pressure). There is a difference of 100 K between the freezing point and boiling point of water. The other widely used temperature scale in science is the Celsius scale, which has units of degree Celsius, °C. In this scale ice melts at 0°C and water boils at 100°C so there is a difference of 100°C, between these phase changes, and therefore the kelvin and celsius units are the same size. We will use both units in different contexts in this book. To convert K to °C, subtract 273.15, and to convert °C to K add 273.15. You may also come across the Farenheit scale in everyday life, but it is not used in science. In this scale, ice melts at 32°F and water boils at 212°F.

The derived SI unit of volume is the cubic metre, m^3. However, this quantity is not a convenient measure in the laboratory, and therefore volumes are either given in cubic decimetres, dm^3, or cubic centimetres, cm^3. The cubic decimetre is exactly the same volume as a litre, L, and therefore dm^3 and L are used more or less interchangeably. In this book we will tend to use L as this is more widely used in environmental science than dm^3. There are 10 dm in 1 m, and therefore there are $10 \times 10 \times 10$ dm^3 in 1 m^3, i.e., 1000 dm^3 or 1000 L in 1 m^3. The cubic centimetre is exactly the same volume as the millilitre, mL, and therefore cm^3 and mL are also used interchangeably. There are 10 cm in 1 dm, and therefore there are $10 \times 10 \times 10$ cm^3 in 1 dm^3, i.e., 1000 cm^3 in 1 dm^3 or 1000 mL in 1 L.

The SI unit of pressure is the pascal, Pa, but two other units are also widely used: the bar, B, and the atmosphere, atm. TV and radio weather forecasters tend to give pressures in millibar, mB, where 1000 mB = 1 B, which is exactly equivalent to 10^5 Pa, or 100 kPa. Standard atmospheric pressure is 1.01325×10^5 Pa or 101.325 kPa, and this has the special unit atm, so that 1 atm = 1.01325×10^5 Pa. We will tend to use either Pa or atm, depending on context in this book. (You may also see references to the torr (which is 1/760 atm) in older texts, but this is uncommon nowadays.)

The unit of atomic weight (or atomic mass) or molecular weight is called the atomic mass unit, amu, and is exactly 1/12th the mass of one ^{12}C atom (as mentioned in Section 1.1). The dalton, Da, where 1 Da = 1 amu, is increasingly being used instead of amu. The atomic weights of the elements are given on the Periodic Table, and also in many tables of masses. The molecular weight of a compound is simply calculated by adding up the atomic weights of all the atoms in the molecular formula. For example, the molecular weight of sulfuric acid, H_2SO_4 is worked out by adding the atomic weights of hydrogen (1.01 amu, to 2 decimal places), sulfur (32.07 amu), and oxygen (16.00 amu), multiplying each atomic weight by the number of atoms of that element in the formula:

$$\text{molecular weight of } H_2SO_4 = (2 \times 1.01) + 32.07 + (4 \times 16.00) \text{ amu} = 98.09 \text{ amu}$$

Atomic or molecular masses are more usually given as the **relative atomic mass**, RAM or **relative molecular mass**, RMM, in which case then there are no units as this is a value *relative* to the mass of 1/12th the mass of a ^{12}C atom.

The SI unit of amount of substance is the mole, mol. One mole is 6.022×10^{23} particles (usually atoms, ions, or molecules), and is the number of carbon atoms in exactly 12.000 g of carbon-12. This number is also known as the Avogadro number. The **molar mass** of an element or compound is the mass of one mole (i.e., 6.022×10^{23} particles) of that element or compound and has units of grams per mole, g mol^{-1}. More simply, it is the mass, expressed in grams, of the atomic or molecular weight. The molar mass of H_2SO_4 for example is 98.09 g mol^{-1}. In the case of elements that exist as molecules, for example, hydrogen, which exists as H_2 molecules, or oxygen, which exists mostly as O_2 molecules but also as O_3 (ozone) molecules, you have to be clear whether you are giving the mass of one mole of *atoms* of the element, or one mole of *molecules* of the element. For example, the molar mass of atomic oxygen is 16 g mol^{-1}, whereas the molar mass of molecular O_2 is 32 g mol^{-1} and the molar mass of ozone, O_3, is 48 g mol^{-1} (all to two significant figures).

The number of moles of a substance can be calculated as shown in Equation 1.25, by dividing the mass of substance (in grams, g) by the molar mass:

$$\text{number of moles of substance} = \frac{\text{mass of substance}}{\text{molar mass}} \qquad (1.25)$$

SELF-ASSESSMENT QUESTIONS

Q1.10 (i)   Calculate the RMM of ammonium nitrate, NH_4NO_3, quoting the relative atomic weights to two decimal places.
      (ii) Express the molecular weight of NH_4NO_3 as a molar mass.
Q 1.11 (i) How many moles are there in 140.12 g of NH_4NO_3?
      (ii) How many moles are there in 3.503 g of NH_4NO_3?

1.3.2 SCIENTIFIC NOTATION AND SI PREFIXES

You have already had a reminder in Box 1.1 of how scientific notation is used to simplify the handling of very large and very small numbers by expressing them as powers of 10. Numbers can also be simplified, however, by changing the units themselves. In this case a prefix is added to the unit to indicate the number of powers of 10 larger or smaller the modified unit is to the original unit. These prefixes are given in Table 1.7. For example, a microgram, μg, is 10^{-6} g, i.e., 1/1000000th of a gram. If you have a measurement of 0.000067 g, you may express this as 6.7×10^{-5} g, or you may prefer to express this as 67 μg. A kilometre, km, is 10^3 m, or 1000 m. If you have a measurement of 540,000 m, you may express this as 5.4×10^5 m, or you may prefer to express this as 540 km. The following examples will illustrate how these prefixes work:

TABLE 1.7
SI Prefixes

Prefix	Symbol	Factor
femto	f	10^{-15}
pico	p	10^{-12}
nano	n	10^{-9}
micro	μ	10^{-6}
milli	m	10^{-3}
centi	c	10^{-2}
deci	d	10^{-1}
deka	da	10
hecto	h	10^2
kilo	k	10^3
mega	M	10^6
giga	G	10^9
tera	T	10^{12}

(i) $0.0934 \text{ g} = 93.4 \times 10^{-3}$ g $(9.34 \times 10^{-2}$ g in scientific notation$) = 93.4$ mg
(ii) $0.00000035 \text{ m} = 0.35 \times 10^{-6}$ m $(3.5 \times 10^{-7}$ m in scientific notation$) = 0.35$ μm
or $0.00000035 \text{ m} = 350 \times 10^{-9}$ m $= 350$ nm
(iii) $438,000 \text{ Pa} = 438 \times 10^3$ Pa $(4.38 \times 10^5$ Pa in scientific notation$) = 438$ kPa
(iv) $62,500,000 \text{ J} = 62.5 \times 10^6$ J $(6.25 \times 10^7$ J in scientific notation$) = 62.5$ MJ

To use a prefix that multiplies the unit by a negative power of 10 as in examples (i) and (ii), move the decimal point the appropriate number of places to the right. To use a prefix that multiplies the unit by a positive power of 10 as in examples (iii) and (iv), move the decimal point the appropriate number of places to the left.

SELF-ASSESSMENT QUESTION

Q1.12 Select appropriate unit prefixes for the following numbers:
(i) 12860 m
(ii) 6830000 J
(iii) 0.00000374 g
(iv) 0.0292 L

1.3.3 CONCENTRATIONS AND SOLUTIONS

Throughout this book we will be referring to the concentrations of different substances in various media, for example, the concentrations of ions in seawater or gases in the atmosphere. The units of concentration therefore require a measure of the amount of substance of interest, and a measure of the total amount of **matrix** (i.e., air, water, rock, etc.) that it is in. For concentrations of aqueous solutions, the main unit of concentration is moles per cubic decimetre, mol dm^{-3}, which is exactly the same as moles per litre, mol L^{-1}, and which is given the symbol M (pronounced *molar*). So, a 2 mol L^{-1} or 2 M solution of H_2SO_4 has 2 moles of H_2SO_4 for every litre of solution. This concentration may also be written as 2 mol dm^{-3}. You should note here that the volume is the volume of the *solution* (i.e., acid + water), not the volume of water by itself. If 2 moles of H_2SO_4 were added to 1 litre of water, the total volume would be in excess of 1 litre. A shorthand way of indicating concentration is to use square brackets around species; for example, [A] means "the concentration of A." The concentration of a solution in moles per litre is calculated as shown in Equation 1.26, where the mass of the dissolved substance is measured in grams and the volume of solution is measured in millilitres, mL, or cubic centimetres, cm^3, which is the same volume.

$$\text{Concentration} = \frac{\text{mass of substance} \times 1000}{\text{molar mass of substance} \times \text{volume of solution}} \text{M} \quad (1.26)$$

SELF-ASSESSMENT QUESTION

Q1.13 Calculate the concentration of a solution of magnesium carbonate, $MgSO_4$, which has 5.32 g of $MgSO_4$ dissolved in 100 mL of water.

In a solid or solution, the amount of substance may be given as the mass of substance per mass of sample, typically, as milligrams of substance per kilogram of sample, mg kg^{-1}, or micrograms of substance per gram of sample, μg g^{-1}. Because there are 1 million mg in 1 kg and 1 million μg in 1 g, these units can also be written as **parts per million**, ppm. Also, 1 L of water has a mass of 1 kg, so for aqueous solutions of *low* concentration, a concentration given in mg L^{-1} is equivalent to mg kg^{-1} or ppm. When using units of ppm with some species, particularly nitrate, NO_3^- and phosphate, PO_4^{3-}, there is potential for confusion as the concentrations of these species are sometimes given as ppm of the anion (the negatively charged species NO_3^- or PO_4^{3-}) and sometimes as the element (N or P), and therefore you need to be clear which is being used. For example, a solution that has a concentration of nitrate as NO_3^- of 31 ppm may also be marked as having a concentration of nitrate as N of 7 ppm (i.e., 31 ppm × 14/62, where 14 is the RAM of N and 62 is the RMM of NO_3^-). For very small amounts, the measurement may be given as micrograms per kilo, μg kg^{-1} or nanograms per gram, ng g^{-1}, which can also be written as **parts per billion**, ppb.

For solid samples in which the concentration of the substance is in excess of about 1000 ppm, concentrations are generally given as percentages. For example, the concentration of silica, SiO_2, in a rock may be 52%; i.e., there is 52 g of SiO_2 per 100 g of rock. To convert from ppm to percentage, divide by 10^4, and to convert from percentage to ppm, multiply by 10^4.

In a gas, concentrations are given as percentage by volume or as parts per million by volume, ppmv, where ppmv is microlitres of substance per litre of gas, or litre of substance per million litres of gas. The concentration of nitrogen in the Earth's atmosphere is 78%, i.e., there is 0.78 L (780 mL) of nitrogen in every litre of air. The concentration of CO_2 in the atmosphere is currently about 382 ppmv; i.e., there is 382 L of CO_2 in every million litres of air.

Gas concentrations are also often given as **partial pressures**, which is the pressure of the gas in the atmosphere expressed as a fraction of the total pressure. This is based on the assumption that the fraction of the total atmospheric pressure being exerted by a particular gas is the same as the fraction of the total volume of the atmosphere occupied by that gas. So, for example, the partial pressure of nitrogen, p_{N2}, is 0.78 atm as this is the fraction of the volume of the Earth's atmosphere occupied by nitrogen (i.e., 78%/100%), whereas the partial pressure of CO_2, p_{CO2}, is 382×10^{-6} atm (i.e., 382 ppmv/10^6). Partial pressure has units of atm, Pa or B. We will be using atm.

You will often see the solidus, or forward slash, /, used in units instead of negative exponents (powers of 10) to indicate that units are being divided. For example, the concentration of a 2 M solution may be given as 2 mol/L rather than 2 mol L^{-1}, or the molar molecular weight of a substance may be given as 96 g/mol instead of 96 g mol^{-1}. In these situations, it is clear that we are dealing with "moles per litre" or "grams per mole". But what if we have a number where the units are given as J/g.°C (the units of specific heat capacity—see Section 3.1.3)? Does this mean J/(g × °C), in which case °C is below the line, or J × °C/g in which case °C is above the line? It is ambiguous. If, however, the units were given as J g^{-1} $°C^{-1}$, it would be clear that we have units of "joules per gram per degree C," and that °C is below the line. Therefore, the use of the solidus is discouraged in units, especially where there are more than two different units. Instead, each unit should have its own exponent.

Box 1.2 Significant Figures, Decimal Places, and Rounding

How accurate can we be in expressing numbers? This depends on how accurate our measurements are. You might, for example, measure a pebble as 3.2 cm long, or the volume of a solution as 432 mL, or the mass of a sample as 28.36 g. The atomic weight of calcium (Ca) in the periodic table in this book is given as 40.078 Da. For the pebble measuring 3.2 cm, you are stating that the length is greater than 3.1 cm but smaller than 3.3 cm and, in fact, that it is nearer to 3.2 cm than to either 3.1 cm or 3.3 cm. This value is quoted to two *significant figures*, i.e., we know to two digits what the length is. The volume of the solution that is given as 432 mL is quoted to three significant figures: we know to three digits what the volume is, and we know that it lies nearer to 432 mL than to either 431 mL or 433 mL. The mass of the sample is given to four significant figures, and we know that it lies nearer to 28.36 g than to either 28.35 g or 28.37 g. The atomic weight of calcium is quoted to five significant figures, which tells us that the value lies nearer to 40.078 Da than to 40.077 Da or to 40.079 Da.

When working out the number of significant figures of a number that has no decimal point, all trailing zeros, that is all zeros to the *right* of all other digits are not significant (i.e., they are *insignificant*). For example, the number 24,000 is given to two significant figures: the three zeros to the right of the 4 are insignificant. However, the number 20,400 is given to three significant figures: the two zeros to the right of the 4 are insignificant, but the 0 between the 2 and 4 is significant.

For numbers smaller than 1, any zeros immediately to the right of the decimal place and to the left of other digits are not significant. For example, the number 0.00246 is given to three significant figures because the two zeros between the decimal place and the 2 are insignificant. Importantly, when zeros appear to the right of other digits after a decimal point they are significant. For example, the number 0.0002300 is given to four significant figures, because the two zeros to the right of the 3 are significant.

In order to be clear how many significant figures you are quoting, scientific notation should be used. For example, consider the mass of a sample that is given as 230 g. As written, this value is given to two significant figures, and indicates that the true mass of the sample is nearer to 230 than to either 220 g or 240 g. If, however, the mass of the sample is written as 2.30×10^2 g, this indicates that the mass is known to three significant figures, and the true mass is nearer to 2.30×10^2 g than to 2.29×10^2 or to 2.31×10^2 g; i.e., it is closer to 230 g than to 229 g or 231 g.

It is frequently necessary to round numbers up or down. For example the displays of most pocket calculators display values to eight or so significant figures, but if the original values used were only given to, say, three significant figures, it is nonsensical to report an answer to eight significant figures. You need to round your answer to an appropriate number of significant figures, and to drop the insignificant figures. If the first digit of the insignificant figures is 5 or above, the last digit of the significant figure is rounded up. For example, the number 243.689 rounded to four significant figures is 243.7, as the 6 will round up to 7 because it is followed by 8. The number 64.97 rounded to three significant figures is 65.0 as the 9 is rounded up to 0, changing the 4 to a 5. If the first digit of the insignificant figures is 4 or below, the last digit of the significant figure is left unchanged. For example, the number 856.234 rounded to four significant figures is 856.2.

You cannot quote a result to more significant figures than appeared in the original data. For example, if you perform a calculation in which the mass of a sample is given to three significant figures as 3.25 g, you cannot report the answer of your calculation to more than three significant figures. A problem comes, however, if you round up your numbers during a calculation. This can lead to significant rounding errors. A useful guide to avoid this is to carry through calculations to one more significant figure than you will quote at the end.

Numbers may also be described according to the number of figures that are shown to the right of the decimal place. The number 3.2 is given to one decimal place, the number 24.85 is given to two decimal places, and the number 40.078 is given to three decimal places.

1.4 CHEMICAL BONDING

You learnt about atoms in Section 1.1. When two or more atoms combine by forming a chemical bond, they form **molecules**. In a few molecules, all the atoms are of the same element. For example, the oxygen in the Earth's atmosphere is mostly in the form of O_2, sometimes called **dioxygen**, in which two oxygen atoms are joined together, whereas a small (but very important) amount of oxygen is present as O_3, ozone, in which three oxygen atoms are joined together. Usually, however, molecules are formed from atoms of two or more different elements. Substances in which atoms from at least two different elements are chemically joined together are called compounds, and the smallest particle of a compound that still retains the identity of the compound is a molecule.

The atoms in molecules are held together by **chemical bonds**. But what is the nature of these chemical bonds? Broadly speaking, they can be grouped into three types: covalent bonds, ionic bonds, and metallic bonds. In covalent bonds, the bonding is based on the sharing of electrons between (usually) two atoms. In ionic bonds, electrons are transferred completely from one atom to another, forming positively and negatively charged ions, and the bonding is based on the **electrostatic** attraction of oppositely charged ions. In metallic bonds, electrons are **delocalised**; that is, they are not **localised** on an atom or ion or between two atoms joined by a covalent bond but they are shared across all atoms in the structure. The atoms pack together surrounded by a "sea" of electrons.

Each of these types of bonding will be considered in turn. You should be aware, however, that although the bonding in a molecule may be classed as ionic, covalent, or metallic, the situation is not really as simple as this and that there is no clear division between the different types of bonding. Bonding can, for example, be partially ionic and covalent if two atoms share their electrons unequally. You will read more about this in Section 1.4.4

1.4.1 Covalent Bonding

The basic principle behind covalent bonding is the sharing of electrons between atoms. This can be considered at many levels, from a very qualitative picture to a highly mathematical treatment. We will be considering the formation of covalent bonds from two viewpoints: achieving stability by forming molecules in which the atoms have full electron shells, and the formation of bonds by overlap of electron orbitals.

As we discussed in Section 1.1.4, a full shell of electrons is a stable arrangement within an atom. Thus, the noble gases have stable electronic configurations because they do not have any partially filled shells; the shells are either full or empty. One way in which atoms can achieve a stable arrangement is by sharing electrons with other atoms. Consider carbon: its electronic configuration is $1s^2 \, 2s^2 \, 2p^2$, which can also be written [He] $2s^2 \, 2p^2$. It has a full first shell ($1s^2$). It has a partly filled second shell ($2s^2 \, 2p^2$) containing four electrons, which are its outermost or **valence electrons**. To achieve a stable noble gas configuration (often called a **stable octet**), it can either lose four electrons to empty the second shell, or gain four electrons to fill the second shell. To lose four electrons would require too much energy. One way of gaining four electrons is to share four electrons with other atoms. Hydrogen has an electronic configuration $1s^1$, so it has one valence electron. To have either a full or empty first shell, it can either lose one electron to form H^+ (this will be discussed

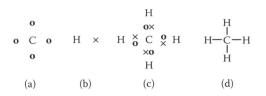

FIGURE 1.17 Lewis (dot-and-cross) diagrams and bonding: Lewis (dot-and-cross) diagrams of (a) carbon, (b) hydrogen, (c) methane. Bonds in methane are shown as single lines to represent shared pairs of electrons.

in Chapter 3, Section 3.3), or it can gain one electron to form H⁻, or it can gain one electron by sharing.

The sharing of electrons can be illustrated by simple dot-and-cross, or **Lewis diagrams**. Figure 1.17a shows carbon with its four valence electrons (given as open circles), and Figure 1.17b shows hydrogen with its one valence electron (shown as a cross). In the molecule methane, CH_4, carbon can gain four electrons by sharing the one electron on each of four hydrogen atoms, and each hydrogen can share one of the electrons on carbon, as shown in Figure 1.17c. Carbon now has eight valence electrons and each hydrogen has two, which is a stable arrangement.

The **bond order** is the number of pairs of electrons shared by two atoms. A bond in which two atoms share one pair of electrons has a bond order of 1, and the bond is called a **single bond**. In methane, each of the four carbon–hydrogen bonds is a single bond. In a chemical structure, single bonds are represented by a single line drawn between the two atoms, as shown in Figure 1.17d. The shared pairs of electrons are not normally shown.

Oxygen has six valence electrons (Figure 1.18a) and therefore requires two more for a stable octet. In carbon dioxide, CO_2, each oxygen atom shares two electrons from the carbon atom, making a stable octet. Similarly, the carbon atom shares two electrons from each oxygen, making a stable octet. Each oxygen is therefore sharing four electrons in total with the carbon, as shown in Figure 1.18b. A bond in which two atoms share four electrons (i.e., two pairs) has a bond order of 2, and the bond is called a **double bond**. In CO_2, there are two double bonds, each oxygen forming a double bond to the carbon. In chemical structures, double bonds are represented by two lines joining the atoms (Figure 1.18c). The electrons not involved in the bonds are drawn in pairs and are called **lone pairs** (Figure 1.18b). They are only shown when it is necessary, otherwise they are omitted.

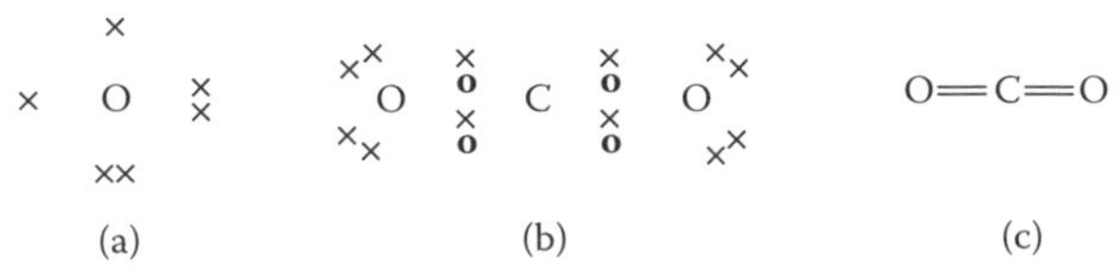

```
    ×                 ×                          
×   O   ×        ×× O  ×    × ×                 
        ×             ×  C  × O ××         O═C═O
    ××            × ×    × ×    ×            
                      ×   O  ×                  
   (a)                 (b)                  (c)
```

FIGURE 1.18 Bonding in carbon dioxide: (a) Lewis diagram of oxygen; (b) Lewis diagram of carbon dioxide, including the two lone pairs on each of the two oxygen atoms; (c) the double bonds shown as two lines between C and O.

Two atoms sharing six electrons (three pairs) have a bond order of 3, and are joined by a **triple bond**, shown by three lines. SAQ 1.15 asks you to draw a Lewis diagram of a molecule with a triple bond.

It is very important that you understand that in all these diagrams, although electrons are marked as crosses or open circles, they are actually indistinguishable from each other. The different symbols used to represent the electrons are only used to aid clarity in the process of counting electrons.

SELF-ASSESSMENT QUESTIONS

Q1.14 Draw Lewis (dot-and-cross) diagrams for:
(i)   hydrogen chloride, HCl
(ii) O_2

Q1.15 The nitrogen molecule N_2 (sometimes called **dinitrogen**) has a triple bond. Draw a Lewis diagram of N_2.

Q1.16 Draw a Lewis diagram for nitrogen dioxide, NO_2. What do you notice? Note: the nitrogen atom is in the middle, with an oxygen on each side.

Covalent bonds vary in their strength, but they are generally strong, and they require a lot of energy to be broken. A typical single bond has a bond energy in the range 250–400 kJ mol^{-1}. Some examples of bond energies are given in Table 1.8. As you can see, the higher the bond order, the stronger the bond, so that, for example, a C-O bond, which has a bond order of 1, is weaker than a C=O bond, which has a bond order of 2. The O=O bond is likewise stronger than the O-O single bond, which is rather weak and easily broken. Molecules, called **peroxides**, which contain an O-O bond, are explosive.

Covalent bonds are actually formed by the overlap of electron orbitals, which enables the sharing of electrons. You will now see how this occurs. You have already seen diagrams of s, p, and d orbitals in Figure 1.5. Some of the ways that they can overlap are shown in Figure 1.19. An orbital can overlap with another of the same type, or an orbital of a different type. New **molecular orbitals** are formed containing the shared electrons.

TABLE 1.8
Bond Energies of Some Typical Bonds

Bond	Bond energy/kJ mol^{-1}	Bond	Bond energy/kJ mol^{-1}
H–H	436	C–O	351
C–H	414	C=O	745
O–H	464	C=O (as in CO_2)	803
C–C	347	C≡O (as in CO)	1075
C=C	611	O–O	130
C≡C	837	O=O (O_2)	498

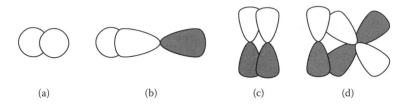

FIGURE 1.19 Overlaps of atomic orbitals to form chemical bonds: (a) overlap of two s orbitals, (b) overlap of an s orbital and a p orbital end on, (c) overlap of two p orbitals sideways on, (d) overlap of a d orbital and a p orbital sideways on.

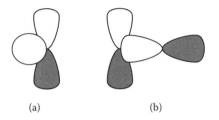

FIGURE 1.20 Overlap of atomic orbitals that will not result in bond formation: (a) overlap of an s orbital and a p orbital sideways on, (b) overlap of a p orbital sideways on with a p orbital end on.

Not every arrangement of overlapping orbitals leads to a bond being formed. This can be illustrated by considering the shading on the orbitals. The diagrams show the s orbital as white, the p orbital with one grey and one white lobe, and the d orbital with alternating grey and white lobes. A bond will only be formed when the overlap is between white and white or between grey and grey, as in the examples of Figure 1.19. Two examples are shown in Figure 1.20 in which overlap will not lead to bond formation, as the overlap is between white and grey+white or grey and grey+white. In some textbooks the shading is replaced by + and − signs, but these do not represent electronic charge. When the + and − convention is used, a bond will only be formed where the overlap is between + and + or between − and −.

SELF-ASSESSMENT QUESTIONS

Q1.17 Sketch a diagram to show two d orbitals overlapping to form a bond.

Q1.18 Sketch a diagram to show a second way that two p orbitals may overlap to form a bond.

Q1.19 Sketch a diagram to show an arrangement of an s orbital overlapping with a d orbital which will not give a bond.

Covalent compounds can be represented by their **molecular formula**, which is the actual numbers of atoms of each element in a molecule, for example, H_2O (water) or C_2H_6O (ethanol); or by an **empirical formula**, which is the simplest ratio of the numbers of atoms of each element in the compound, for example, CH_2O (this is the empirical formula for cellulose, which forms the bulk of plant material); or by a **condensed formula**, which shows how the atoms are grouped together, for example, C_2H_5OH or CH_3CH_2OH (both are also ethanol); or by a **structural formula**, which shows the bonds between the atoms. We will be considering the structures of various kinds of covalent compounds in Section 1.5.1.

1.4.2 CATIONS, ANIONS, AND IONIC BONDING

The basic principle of ionic bonding is the **electrostatic** attraction between oppositely charged ions. An atom or molecule that loses one or more electrons forms a positively charged **cation**, and an atom or molecule that gains one or more electrons forms a negatively charged **anion**. To be able to discuss ionic bonding and ionic compounds, it is first necessary to be familiar with the common cations and anions.

The metals of group 1 (the alkali metals) have an electronic configuration of [noble gas] ns^1. They all form ions with a single positive charge, M^+, because the loss of the single s electron in their outermost orbital will leave them with only full or empty electron orbitals, which is a stable arrangement. The metals of group 2 have an electronic configuration of [noble gas] ns^2. They all form ions with a +2 charge, M^{2+}, because the loss of the two s electrons in their outermost orbital will leave them with only full or empty electron orbitals. Aluminium in group 13 has an electronic configuration of [Ne] $3s^2\ 3p^1$, and it forms a cation with a +3 charge, Al^{3+}, owing to the loss of the three electrons in its outermost orbital. The transition metals of groups 3–12 form ions of various charges. For example, iron, Fe, in group 8 forms two common cations, Fe^{2+} and Fe^{3+}, by losing, respectively, two or three electrons from the atom. Examples of some common cations are given in Table 1.9.

Oxygen and sulfur in group 16 have an electronic configuration of [noble gas] $ns^2\ np^4$. They form anions with a –2 charge, X^{2-}, because the gain of two electrons will fill their outermost electron orbital. The halogens of group 17 have an electronic configuration of [noble gas] $ns^2\ np^5$, and they all form anions with a single negative charge, X^-, because the gain of a single electron will fill their outermost electron orbitals. The names of these anions end in "–ide," e.g., oxide, and examples of some common anions are given in Table 1.10. Lewis dot-and-cross diagrams of calcium, the Ca^{2+} cation, oxygen, and the O^{2-} anion are shown in Figure 1.21. As you can see, two electrons are removed from the Lewis diagram of calcium to give the +2 charge on the Ca^{2+} cation, whereas two electrons are added to the Lewis diagram of oxygen to give the –2 charge on the O^{2-} anion.

TABLE 1.9
Some Common Cations

Cation	Example
M^+	H^+, Li^+, Na^+, K^+, Rb^+
M^{2+}	Mg^{2+}, Ca^{2+}, Sr^{2+}, Ba^{2+}, Mn^{2+}, Fe^{2+}, Co^{2+}, Cu^{2+}, Ni^{2+}, Zn^{2+}
M^{3+}	Al^{3+}, Cr^{3+}, Fe^{3+}
M^{4+}	Ti^{4+}, Si^{4+}

TABLE 1.10
Some Common Anions

Anion Name	Anion Formula
Fluoride	F^-
Chloride	Cl^-
Bromide	Br^-
Iodide	I^-
Hydroxide	OH^-
Oxide	O^{2-}
Sulfide	S^{2-}

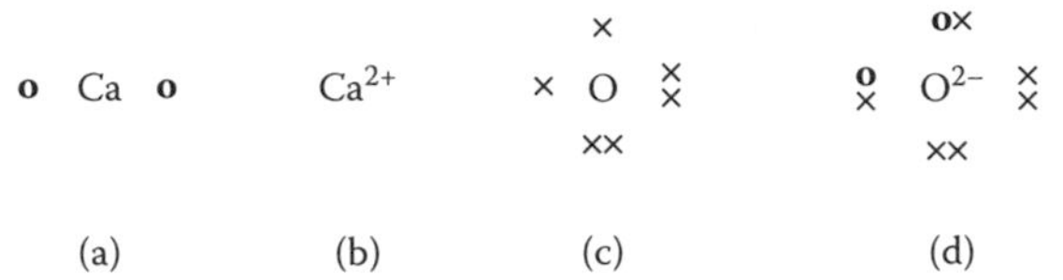

o Ca o Ca^{2+} × O ×   ×O O^{2-}×

(a)   (b)   (c)   (d)

FIGURE 1.21 Lewis diagrams of calcium and oxygen showing: (a) the two valence electrons of a calcium atom, (b) that Ca^{2+} has an empty outer shell, (c) the six valence electrons of an oxygen atom, (d) that O^{2-} has a full electron shell.

Nonmetals of groups 14–16, e.g., carbon, C, nitrogen, N, phosphorus, P, and sulfur, S, form a range of anions with oxygen called **oxoanions**. Some of these anions occur naturally in the environment in rocks and minerals, e.g., carbonate, CO_3^{2-} and sulfate, SO_4^{2-}, whereas some occur partly as a result of pollution, e.g., nitrate, NO_3^-, from agricultural run-off, and phosphate, PO_4^{3-}, from agriculture and detergents. Some of the more common oxoanions are given in Table 1.11. The Lewis diagram for nitrite, NO_2^-, is shown in Figure 1.22. Nitrogen has five valence electrons ($2s^2\ 2p^3$) and oxygen has six ($2s^2\ 2p^4$). The nitrogen electrons are the crosses, the oxygen electrons are the open circles, and the extra electron forming the anion is the filled circle.

TABLE 1.11
Some Common Oxoanions

Anion Name	Anion Formula
Carbonate	CO_3^{2-}
Bicarbonate (hydrogen carbonate)	HCO_3^-
Nitrate	NO_3^-
Nitrite	NO_2^-
Phosphate	PO_4^{3-}
Sulfate	SO_4^{2-}

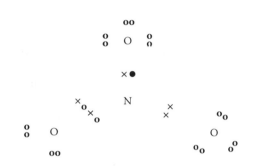

FIGURE 1.22 Lewis diagram of Nitrite, NO_2^-.

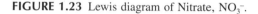

FIGURE 1.23 Lewis diagram of Nitrate, NO_3^-.

It is possible for an atom to share its lone pairs with another atom to make up the full shell of the second atom. This is what happens when nitrite is oxidised (gains an oxygen atom) to form nitrate. The Lewis structure for NO_3^- is shown in Figure 1.23, and as you can see, the third oxygen atom achieves a filled shell using the lone pair of electrons from the nitrogen.

SELF-ASSESSMENT QUESTION

Q1.20 Draw the Lewis structure for carbonate, CO_3^{2-}.

Many simple compounds are ionic, and in these compounds the overall charge provided by cations and anions must balance. The simplest ionic compound is one in which a cation with a single +1 charge is balanced by an anion with a single –1 charge. A well-known example is sodium chloride (common salt), NaCl, also known as the mineral halite, which consists of Na^+ and Cl^- ions. A cation with a +2 charge can be balanced by one anion with a –2 charge as in calcium carbonate, $CaCO_3$, which is formed from calcium, Ca^{2+}, and carbonate, CO_3^{2-} ions, or two anions each with a –1 charge as in calcium fluoride, CaF_2, which is formed from Ca^{2+} and fluoride, F^-, ions. Similarly, an anion with a –2 charge such as CO_3^{2-}, can be balanced by two cations with a +1 charge, as in sodium carbonate, Na_2CO_3, or a cation with a +2 charge as we have already mentioned for $CaCO_3$. In ferric sulfate, $Fe_2(SO_4)_3$, two Fe^{3+} (iron(III) or ferric) ions are balanced by three SO_4^{2-} (sulfate) ions. In the mineral boehmite, $AlO(OH)$, the Al^{3+} cation is balanced by one oxide, O^{2-}, and one hydroxide, OH^-, ion. $AlO(OH)$ may also be written as $AlOOH$.

SELF-ASSESSMENT QUESTIONS

Q1.21   What ions are present in potassium iodide, KI? What are the sizes of the charges on the ions?

Q1.22   Anhydrite, a common mineral in sedimentary rocks, consists of Ca^{2+} and SO_4^{2-} ions. What is its formula?

Q1.23   Brucite, a mineral characteristic of metamorphic rocks, consists of magnesium, Mg^{2+}, and OH^- ions. What is its formula?

1.4.3 METALLIC BONDING

The basic principle of metallic bonding is that metal atoms pack together in a "sea" of electrons that are **delocalised** (not located on or between specific atoms). This is illustrated in Figure 1.24, in which atomic nuclei with their inner electrons are represented by the $\oplus$ symbol, and the electrons are represented by the grey shading. The electrons can move around freely, and this gives rise to the property of electrical conductivity, which is a very important feature of metals. The "sea of electrons" model is a very qualitative picture and does not account for many of the properties of a metal, but is suitable enough for this discussion.

At a simplistic level, the strength of the bonding will depend on the number of electrons per atom in the "sea." Electrons in filled shells will not contribute to the sea, only electrons in partially filled shells. Sodium and the other group 1 metals, which have only one electron per atom in the outermost, unfilled shell, are very soft and have low melting points, whereas titanium, which has four valence electrons per atoms, is an exceedingly hard metal. Moving along each row of the periodic table, the increasing number of valence electrons initially makes each succeeding metal harder and have a higher melting point than the previous metal. However, the valence electrons become increasingly **localised** on the atoms, so they do not all contribute to the sea, and there are fewer electrons to contribute to the bonding. This means that the metals further along the rows (nearer the nonmetals) get softer, and have lower melting points. Thus, for example, mercury is a liquid, and pure gold is very soft. Boiling points of all metals are high, however, because for an atom to boil, it has to break free completely of the remaining atoms and the electron sea.

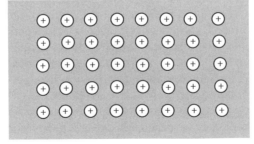

FIGURE 1.24 Metal atoms in a sea of electrons.

Many metals are malleable (can be beaten into sheets) or ductile (can be drawn into wires) because the atoms can easily slide past each other into new positions without breaking the attraction to the electron sea. Gold can be beaten into sheets of remarkable thinness—about 230 atoms thick! This malleability and ductility of metals contrasts with the lack of such behaviour in ionic compounds. This is because the layers of ions in ionic lattices are prevented from sliding over each other by the repulsions between cations and cations and between anions and anions.

1.4.4 ELECTRONEGATIVITY, POLAR BONDS, AND HYDROGEN BONDING

Up to now, we have assumed (without actually saying so) that in covalent bonds electrons are equally shared between two atoms. However, this is not necessarily the case. This is because of **electronegativity**. Electronegativity is the tendency of an atom in a molecule to pull electrons towards itself. When two atoms form a bond, the more **electronegative** atom will take a greater share of the electrons and will become slightly negatively charged (indicated by a δ– symbol), whereas the less electronegative (or more **electropositive**) atom will have a lesser share of the electrons and will become slightly positively charged (indicated by a δ+ symbol). The electron cloud is **polarised** or distorted, i.e., is not shared equally between the atoms forming the bond, and the bond is described as being **polar**. If there is no difference in electronegativity, then there will be no polarisation (distortion) of the electron cloud, and the bond is described as **nonpolar**. Figure 1.25a shows the electrons being shared equally between the two oxygen atoms in the nonpolar O=O bond. Figure 1.25b shows the unequal sharing of electrons between hydrogen and chlorine in the polar H-Cl bond, with the more electronegative chlorine taking a greater share of the electron cloud, and therefore carrying a δ– charge, whereas hydrogen carries a δ+ charge. Figure 1.25c shows the partial charges in water. Oxygen is the more electronegative atom, so it has a greater share of electrons; it therefore carries a δ– charge, whereas the two hydrogen atoms have a lesser share of the electrons, and each carries a δ+ charge.

The greater the difference in electronegativities between two atoms, the greater the degree of polarisation of the bond. As the bond becomes more polar, it becomes more ionic than covalent in character, until (as in the case of NaCl, for example) it

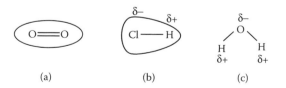

(a)   (b)   (c)

FIGURE 1.25 Polar bonds and the distortion of the electron cloud. (a) equal sharing of the electron cloud in the O-O bond in O_2; (b) unequal sharing of the electron cloud in the H-Cl bond in HCl, showing also partial charges on H and Cl; (c) partial charges on the O and H atoms in H_2O.

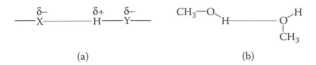

(a)   (b)

FIGURE 1.26 Hydrogen bonding: (a) δ+ hydrogen attracted to δ– nitrogen, oxygen, or fluorine; (b) hydrogen bonding in methanol.

is essentially an ionic bond. Metals have electronegativities less than 2, whereas nonmetals have electronegativities greater than 2. Metalloids have electronegativities of about 2.

Just as cations and anions are attracted to each other by electrostatic attraction, the δ+ and δ– ends of molecules are also attracted to one another. The attraction between δ+ and δ– atoms is particularly marked in bonds where hydrogen is bonded to the most electronegative elements, nitrogen, oxygen, and fluorine. This is illustrated in Figure 1.26a, where X and Y may be any combination of nitrogen, oxygen, and fluorine. When hydrogen is bonded to one of these atoms (Y in Figure 1.26a), it will be strongly δ+, and will therefore be strongly attracted to the δ– charge (in the shape of the lone pairs) on a nitrogen, oxygen, or fluorine atom (X in the diagram) in another molecule or even (if the positions of the atoms is suitable) in the same molecule. The bond formed, which is shown by the ---- marking, is called a **hydrogen bond**. The hydrogen bond formed in the simple organic molecule methanol, CH_3OH, is given in Figure 1.26b as an example. Polar molecules such as hydrogen chloride that cannot form hydrogen bonds because they do not contain O, N, or F atoms are held together by a weaker attraction called a **dipole–dipole** interaction.

Hydrogen bonds are much weaker than single covalent bonds, and have bond energies on average of about 25 kJ mol^{-1}. Nevertheless, although this is considerably less than the energy of a single covalent bond, hydrogen bonding can make an appreciable difference to the properties of a system. One of the most important properties of water is its ability to hydrogen-bond. Extensive hydrogen bonding between water molecules is responsible for water having such a high boiling point for such a small molecule, and a remarkably large temperature range over which it is liquid. It is also responsible for the fact that water expands rather than contracts on freezing. Hydrogen bonding also holds together the two chains that make the double helix of DNA, and it is the mechanism by which DNA is able to replicate itself. You will learn more about hydrogen bonding in water and how it affects water's physical and chemical properties, and the consequences this has for Earth, its climate, and its life, in Chapter 3.

1.5 CHEMICAL STRUCTURES

In Section 1.4 we looked at the different kinds of bonds that can be formed between atoms or ions. We will now consider the structures that different kinds of compounds can form. We will start by distinguishing between **molecular compounds**, in which atoms are joined together to form discrete and well-defined molecules, and **extended networks**, in which atoms or ions are joined together to form structures in which there are no clearly defined molecules. We will further divide molecular compounds into organic compounds and inorganic compounds. Molecular compounds range in size from just two atoms to thousands of atoms. Examples of molecular compounds include carbon dioxide, CO_2, and water, H_2O, with just three atoms each; benzene, C_6H_6, with twelve atoms; and large biomolecules such as enzymes and chromosomes with many thousands. Extended networks consist of arrays of covalently bonded atoms, or lattices of ions held together by ionic bonding. Examples include sodium chloride, NaCl, ferric chloride (rust), Fe_2O_3, and most silicate minerals, as well as the two main polymorphs of carbon, diamond and graphite. We will be considering the structures of silicate minerals in some detail in Chapter 2, Section 2.2.

1.5.1 STRUCTURES OF ORGANIC COMPOUNDS

In this section we will consider the structures of **organic compounds**. Organic compounds are, with a very few exceptions, all compounds that contain carbon. The most important exceptions are carbon monoxide, CO, carbon dioxide, CO_2, and compounds that contain carbonate, CO_3^{2-}, which are considered **inorganic** compounds. In order to be able to understand the structures of molecular compounds, it is necessary to know how many bonds each different type of atom forms. This is known as the **valence** of the atom. The number of bonds made by the elements that commonly occur in organic compounds is given in Table 1.12. When counting the number of bonds, a double bond counts as two bonds and a triple bond counts as three.

Carbon is unique among all the elements of the periodic table in the extent to which it can form chemical bonds to itself as well as other elements, and this gives rise to an enormous number (well over two million) and variety of compounds. To bring some sense of order to the vast range of organic compounds, they are classified according to the **functional groups** they contain. A functional group is a particular atom or arrangement of atoms that occurs in different compounds. The structures of very many organic compounds are based on a chain or chains of carbon atoms, and the length of the longest chain in a molecule (which forms, in effect, the backbone of the molecule) is another way in which organic compounds are classified. The length of a carbon chain is indicated in the compound name by having a different name for each chain length. It is impossible in the space available to us (and it is not necessary) to describe every type of organic compound that exists. We will concentrate on some of the major classes of compounds, and also consider the structures of some of the most important organic pollutants.

Organic compounds that are made from carbon and hydrogen only are called **hydrocarbons**. Hydrocarbons that have only single bonds are called **alkanes** and are **saturated** molecules (i.e., there are no double or triple bonds in the molecules). The molecule name is made by adding **–ane** to the chain name. Alkanes can be **straight chain** (sometimes referred to as **normal** or *n*-alkanes), **branched**, or **cyclic**.

TABLE 1.12

Numbers of Bonds Made by Elements in Organic Compounds

Element	Number of Bonds Made
H, F, Cl, Br, I	1
O, S	2
N	3
C	4

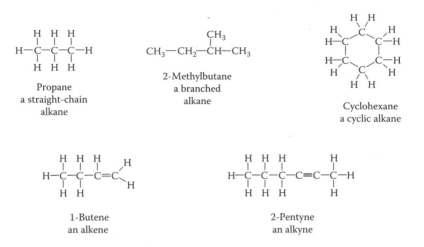

FIGURE 1.27 Structures of alkanes, alkenes, and alkynes.

Branched alkanes have shorter chains attached to the main chain, whereas cyclic alkanes have rings of carbon atoms as shown in Figure 1.27. Straight-chain and branched alkanes have the general formula C_nH_{2n+2}, whereas cyclic alkanes have the general formula C_nH_{2n}. Alkanes are important environmentally as they are a major constituent of crude oil and its derivatives such as petrol (gasoline). Molecules that have one or more C=C double bonds are called **alkenes** (or **olefins** in older literature), and the molecule name is made by adding **–ene** to the chain name. An alkene containing only one C=C bond has the general formula C_nH_{2n}. Alkenes are examples of **unsaturated** molecules (i.e., molecules that have at least one double or triple bond). You have probably heard of the terms "mono-unsaturated" and "poly-unsaturated" applied to margarine. These terms apply to fats that have one, or more than one, C=C double bond, respectively, in their long carbon chains. Molecules that have one or more C≡C triple bonds are called **alkynes**, and the molecule name is made by adding **–yne** to the chain name.

Many organic compounds contain the **halogens** (elements of group 17), fluorine, chlorine, bromine, or iodine. These are known as **halo-compounds**, and some examples are shown in Figure 1.28. Many halo-compounds are of environmental concern because they are responsible for the formation of the hole in the ozone layer in the Earth's atmosphere (see Chapter 4, Section 4.5). These include **chlorofluorocarbons** (CFCs, which contain carbon, fluorine, and chlorine), **hydrochlorofluorocarbons** (HCFCs, which contain carbon, hydrogen, fluorine, and chlorine), and **halons** (which contain bromine, fluorine, carbon, and, sometimes, chlorine and hydrogen). **Hydrofluorocarbons** (HFCs) contain carbon, hydrogen, and fluorine only, and no chlorine or bromine, and therefore do not affect the ozone layer.

There are many different classes of oxygen-containing compounds, but the most important are **alcohols, ethers, aldehydes, ketones, carboxylic acids,** and **esters**. These are shown in Figure 1.29. All of these compounds except alcohols and ethers contain the **carbonyl** group, >C=O, in which oxygen is double-bonded to a carbon atom. The carbonyl compounds are then differentiated according to the other two groups attached to the carbon atom. Alcohols contain the –OH group, and ethers contain

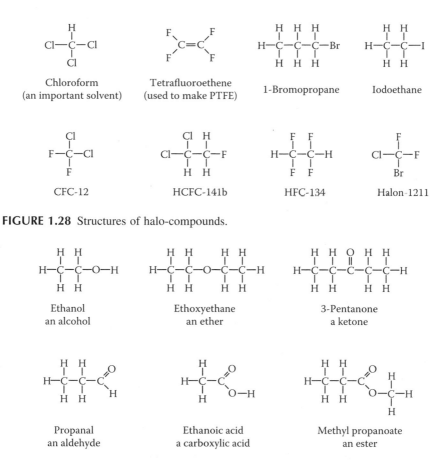

FIGURE 1.28 Structures of halo-compounds.

FIGURE 1.29 Structures of oxygen-containing compounds.

the –O– group. Alcohols, aldehydes, ketones, and acids are named by adding **–anol**, **–anal**, **–anone**, and **–anoic acid**, respectively, to the chain name. Ethanol is the alcohol of alcoholic drinks, but it is also being increasingly used as a renewable biofuel, mixed with petrol. The structure of soil organic matter is very complex (see Chapter 2, Section 2.8), but it contains many of these oxygen-containing functional groups.

Four important classes of nitrogen-containing compounds are **amines** (which contain the –NH₂ group), **amides** (-CO-NH₂), **nitriles** (–C≡N), and **nitro** compounds (–NO₂), and examples are shown in Figure 1.30. Nitrogen is an important constituent of DNA, and amide groups are a fundamental part of the long chains that form protein molecules; therefore, nitrogen is a key nutrient for all organisms. Nitrogen is also a fundamental constituent of the chlorophyll molecule, which controls photosynthesis in plants.

Two important classes of sulfur-containing compounds are **thiols**, which contain the –SH group, and **thioethers**, which contain the –S– group (Figure 1.31). Dimethyl sulfide is formed in the oceans by phytoplankton, and is released into the atmosphere, where it is an important part of the global sulfur cycle (see Chapter 4, Section 4.3.3).

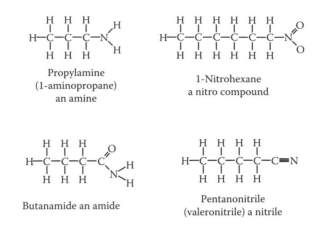

Propylamine
(1-aminopropane)
an amine

1-Nitrohexane
a nitro compound

Butanamide an amide

Pentanonitrile
(valeronitrile) a nitrile

FIGURE 1.30 Structures of nitrogen-containing compounds.

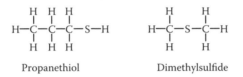

Propanethiol

Dimethylsulfide

FIGURE 1.31 Structures of sulfur-containing molecules.

Many compounds contain the **benzene** ring. These are known as **aromatic** compounds. Benzene itself has the formula C_6H_6. Other functional groups or carbon chains can be added to the benzene ring by replacing hydrogen atoms. For example, a methyl ($-CH_3$) group can be added to benzene to form methylbenzene (also known as toluene), a carboxylic acid ($-COOH$) group to form benzoic acid, or a hydroxyl ($-OH$) group to form phenol (Figure 1.32). Aromatic compounds may contain more

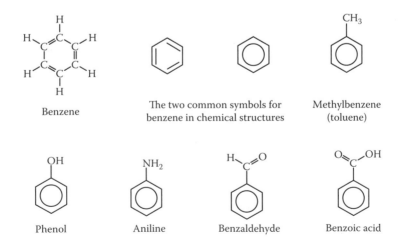

Benzene

The two common symbols for
benzene in chemical structures

Methylbenzene
(toluene)

Phenol   Aniline   Benzaldehyde   Benzoic acid

FIGURE 1.32 Structures of aromatic compounds.

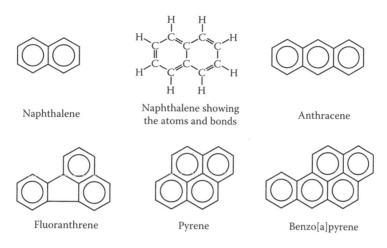

FIGURE 1.33 Structures of polyaromatic hydrocarbons (PAHs).

than one substituent on the benzene ring. Benzene itself and methylbenzene (toluene) are constituents of crude oil, and also (depending on the source) of petrol (gasoline). Benzene is widely used as an intermediate in the manufacture of many chemicals. It is of particular concern in the environment because it is strongly suspected of causing cancer, and because it is volatile (it has a high vapour pressure and evaporates readily), adding to air pollution.

Benzene rings can be joined together. An important class of such compounds is the **polyaromatic hydrocarbons** (PAHs), in which two or more rings are fused by sharing edges (Figure 1.33). PAHs are serious pollutants and are designated as **persistent organic pollutants** (POPs) by the United Nations (UN). Biphenyl is a compound in which two benzene rings are joined by a single bond through two carbon atoms, and it forms the basis of the **polychlorinated biphenyls** (PCBs), in which some or all of the hydrogen atoms are replaced by chlorine atoms. The structures of biphenyl and two example PCBs are shown in Figure 1.34. There are, in fact, 209 PCBs based on the total number of possible arrangements of chlorine atoms that can be attached to the biphenyl backbone. These compounds are entirely synthetic, and like PAHs, PCBs are designated as POPs.

Compounds that have the same molecular formula but different structures are called **structural isomers**. These are formed by bonding atoms in different arrangements. Two structural isomers having the molecular formula C_2H_6O are shown in Figure 1.35a (ethanol) and Figure 1.35c (diethyl ether). These are structural isomers because the atoms are bonded differently. The structure in Figure 1.35b is also ethanol, because although the atoms are drawn differently on the page to Figure 1.35a, they are nevertheless bonded to the same atoms.

SELF-ASSESSMENT QUESTION

Q1.24 Draw as many different structural isomers as you can for (i) C_3H_8O and (ii) C_3H_9N.

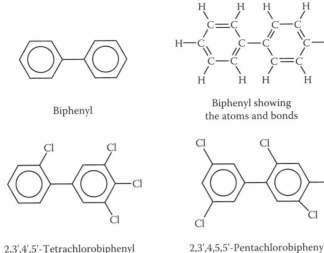

Biphenyl

Biphenyl showing
the atoms and bonds

2,3',4',5'-Tetrachlorobiphenyl

2,3',4,5,5'-Pentachlorobiphenyl

FIGURE 1.34 Structures of polychlorinated aromatic compounds.

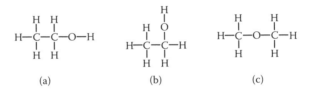

(a)   (b)   (c)

FIGURE 1.35 Isomers of C_2H_6O: (a) Ethanol. (b) This structure is also ethanol, as the same atoms are bonded to each other, just written differently on the page. (c) Dimethyl ether. The atoms are bonded differently to the atoms in ethanol, and therefore this is an isomer of ethanol.

1.5.2 THREE-DIMENSIONAL STRUCTURES, CHIRALITY, AND OPTICAL ISOMERS

When we show chemical structures on paper, we need to be aware that we are representing a three-dimensional structure in two dimensions. Some of the structures that you have seen in Figures 1.27–1.35 are flat, such as ethene and benzene, and so this is a reasonable representation of their three-dimensional structure, but for most molecules we are showing how the atoms are bonded to each other, but not their arrangement in three-dimensional space. A carbon atom that is bonded to four separate groups will form four single bonds. Each pair of electrons in the four bonds is repelled by the other pairs of electrons (remember that like charges repel), so that the four bonds keep as far apart from each other as possible. This means that the four bonds point to the corners of a tetrahedron (triangle-based pyramid), with the carbon atom at the centre, and the angle between each of the bonds is 109.5°, as shown in Figure 1.36. The bonds shown as a normal line lie in the plane of the paper, whereas the bonds shown as arrow-shaped, i.e., ◄ or ►, enter or leave the plane of the paper. The wide end of the arrow is the end that is closer to the reader.

A carbon atom that has four *different* groups attached to it is **chiral**. This means that there are actually two different arrangements in space of the four groups around the central carbon. These two arrangements are called **optical isomers**. They are mirror images of each other, but the mirror images are not superimposable. You can use your own hands as an illustration of chirality: your left and right hands are mirror images of each other,

FIGURE 1.36 The three-dimensional arrangement around carbon.

but you cannot superimpose your hands: you can put the fingers together in the correct order, but your palms are together, so that the palm of one hand would be superimposed on the back of the other hand.

Figure 1.37 shows a chiral carbon atom bonded to four different groups: a methyl group, -CH₃; an amine group, -NH₂; a hydrogen atom, -H; and a carboxylic acid group, -COOH. The molecule is, in fact, alanine, which is one of the amino acids that are the building blocks of proteins in living organisms. The two optical isomers of alanine are called L-alanine and D-alanine. L-alanine is the mirror image of D-alanine. If we turn L-alanine around and try to superimpose it on D-alanine, we find that we cannot. We would have to swap over the H and CH₃ groups to do this. The D and L terminology is used to distinguish the different optical isomers of chiral molecules that occur in nature such as sugars and amino acids. The L form of amino acids is the only form that occurs in proteins.

Many other biologically important molecules are chiral. In virtually every case where chiral molecules occur in biological systems, only one form will be biologically active, and the other form will be inactive. This is because many biological reactions work on the **lock-and-key principle**: when two molecules come together, the atoms must fit together in a three-dimensional arrangement, and only one arrangement of a chiral molecule will fit (Figure 1.38). Only the active form is normally synthesised in nature, and thus nature is said to be handed. There is currently much debate among scientists as to how this handedness of nature arose.

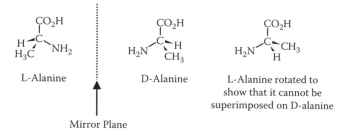

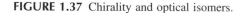

FIGURE 1.37 Chirality and optical isomers.

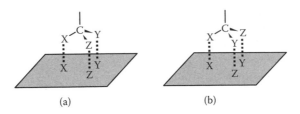

FIGURE 1.38 The lock-and-key principle: (a) the groups on a chiral compound matching those on a substrate, (b) the groups on the optical isomer of the compound in (a) showing that they do not match those on the same substrate.

1.5.3 STRUCTURES OF MOLECULAR INORGANIC COMPOUNDS

Inorganic compounds are compounds of all elements except carbon, but including the handful of carbon compounds not considered as organic (carbon dioxide, carbon monoxide, etc.). There is also a category of compounds called **organometallic** compounds, which have at least one bond between a carbon atom and a metal atom, but we will not consider these here. Three very important (from an environmental perspective) inorganic compounds are water, carbon dioxide, and sulfur dioxide. All of these compounds contain three atoms linked in a line, but whereas CO_2 is a linear molecule, H_2O and SO_2 are bent (Figure 1.39). In CO_2, the two pairs of electrons that form one of the C=O double bonds try to get as far away as possible from the two pairs of electrons that form the other C=O bond (remember that like charges repel), and this is achieved when O=C=O is linear (Figure 1.39a). In H_2O, the pairs of electrons forming the two O–H bonds not only repel each other, but are also repelled by the two lone pairs of electrons on the oxygen atom (Figure 1.39b). This causes the molecule to be bent, with an angle of 104.5° between the two oxygen atoms. The four pairs of electrons (two bonding and two lone pairs) are arranged so that they are almost pointing at the corners of a tetrahedron. In SO_2, there is only one lone pair repelling the electrons that form the two S=O bonds, so the molecule is bent with an angle of 119° between the two oxygen atoms, i.e., the lone pair and the two sets of bonding electrons are pointing towards the corners of an equilateral triangle (Figure 1.39c). (Note: Because sulfur is in the third row of the periodic table, it is no longer restricted to having only eight electrons in its valence orbital.)

Another simple important inorganic molecule is ammonia, NH_3 (Figure 1.39d). This is a pyramidal molecule, rather than triangular, with an angle of 107°, again because the one lone pair of electrons on the nitrogen atom repels the electrons that form the three N-H bonds. By contrast, sulfur trioxide, SO_3 (Figure 1.39f), is a planar molecule. There are no lone pairs, only three double bonds (as noted above, sulfur is not restricted to having eight electrons in its outermost orbital: it can expand its octet), which point to the corners of an equilateral triangle. Both of these molecules form important ions. Ammonia can readily gain a proton to form the ammonium ion, NH_4^+, which is tetrahedral (Figure 1.39e), whereas sulfur trioxide can gain an oxide ion in reaction with water to form sulfate, SO_4^{2-}, which is also tetrahedral (Figure 1.39g).

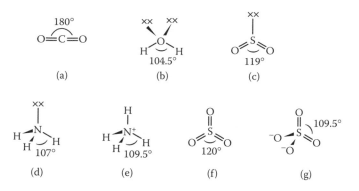

FIGURE 1.39 Shapes of inorganic molecules and ions: (a) CO_2 (linear); (b) H_2O (bent); (c) SO_2 (bent); (d) NH_3 (pyramidal); (e) NH_4^+ (tetrahedral); (e) SO_3 (trigonal); (f) SO_4^{2-} (tetrahedral).

SELF-ASSESSMENT QUESTION

Q1.25 The angle between the two oxygen atoms in NO_2 (whose Lewis dot-and-cross diagram you drew in Q1.16) is 132°. Can you account for this? (Assume that there is one unpaired electron on nitrogen, and all the other electrons are involved in the N-O bonds.)

1.5.4 STRUCTURES OF EXTENDED NETWORKS

As we have already mentioned, extended networks are formed by ionic compounds in which ions are arranged in a regular lattice held together by electrostatic attraction, by covalent compounds, which do not form discrete molecules, and by metals. The structures of compounds that form extended networks are most simply described by considering the ions or atoms as hard spheres that pack together in different arrangements. Ions form close-packed layers, shown in Figure 1.40, which then stack in regular sequences. There are two different common stacking arrangements. In the **cubic close packing** (CCP) or **face-centred cubic**, (FCC) arrangement, the second layer of ions sits in the "dips" of the first layer. The third layer sits in the dips in the second layer but not directly above the first layer. The fourth layer sits in the dips in the third layer, directly above the first layer (Figure 1.41a). If we label the first layer A, the second layer B, and the third C, this gives a layer-stacking sequence of A, B, C, A, B, C, ... The FCC structure is shown in Figure 1.41b, and you can see that the atoms lie at the corners of a cube and in the middle of each face of the cube. The layers actually lie diagonally across the cube, and are high-lighted in grey.

In the **hexagonal close packing** (HCP) arrangement, the first two layers of ions are the same as for the FCC arrangement. The third layer sits in the dips in the second layer, directly over the first layer, giving a stacking sequence of A, B, A, B,... (Figure 1.42a). The HCP structure is shown in Figure 1.42b, and you can see the hexagonal symmetry.

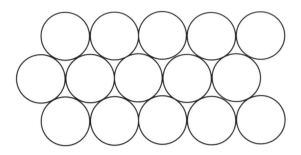

FIGURE 1.40 Close packing of atoms or ions in layers.

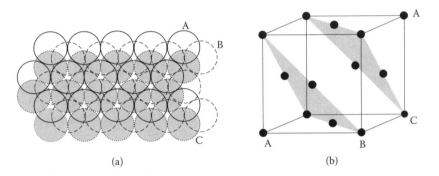

FIGURE 1.41 Face-centred cubic packing showing: (a) the C layer (grey circles) sitting in the dips of the B layer (dashed circles), not directly above the A layer (open circles); (b) the FCC structure with its cubic symmetry. The planes are marked with the grey shading.

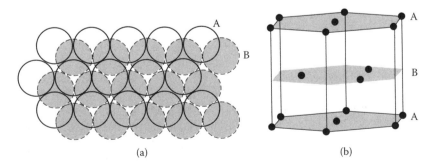

FIGURE 1.42 Hexagonal close packing showing: (a) the B layer (grey circles) sitting in the dips of the A layer (open circles), (b) the HCP structure with its hexagonal symmetry. The planes are marked with the grey shading.

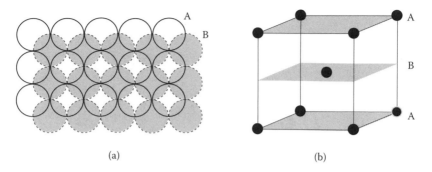

(a)   (b)

FIGURE 1.43 Body-centred cubic packing showing: (a) two square packed layers, with the B layer (grey circles) sitting in the dips of the A layer (open circles); (b) the BCC structure with its cubic symmetry. The planes are marked with the grey shading.

An arrangement common in structures of metals is **body-centered cubic** (BCC). In this arrangement, square-packed arrays of atoms lie on top of each other, with each layer lying in the dips of the layer below it (Figure 1.43a). The BCC structure is shown in Figure 1.43b. Most metals adopt one or other of the FCC, HCP, or BCC arrangements. For example, the group 1 metals all adopt the BCC structure, calcium and aluminium both adopt the FCC structure, and magnesium and zinc adopt the HCP structure.

In both the FCC and HCP arrangements described earlier, there are "holes," or interstices, into which small atoms or ions can fit (Figure 1.44). There are two types of hole: tetrahedral holes, in which an atom or ion in the hole is surrounded by four other atoms or ions (Figure 1.44b), and octahedral holes, in which an atom or ion in the hole is surrounded by six other atoms or ions (Figure 1.44c).

We will now consider the structures important in minerals to illustrate the ways in which close-packing forms extended network structures. The simplest structure is that of sodium chloride, NaCl, which forms the mineral halite. Halite occurs widely in evaporite deposits, which you will learn about in Chapter 2, Section 2.4.3.

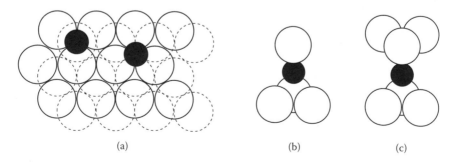

(a)   (b)   (c)

FIGURE 1.44 Tetrahedral and octahedral holes showing: (a) the lower layer (open circles) and upper layer (dashed circles) with atoms or ions (small filled circles) in a tetrahedral hole (left) and in an octahedral hole (right); (b) arrangement of atoms or ions (open circles) around an atom or ion (filled circle) in a tetrahedral hole; (c) arrangement of atoms or ions around an atom or ion in an octahedral hole.

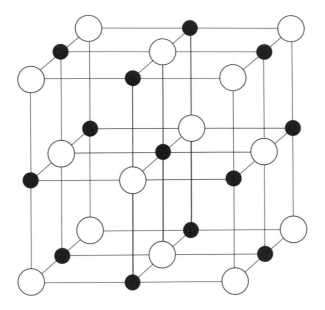

FIGURE 1.45 The structure of sodium chloride.

The structure of halite is important because it is adopted by many ionic compounds with the general formula MX, for example, periclase, MgO (a metamorphic mineral), galena, PbS (the principal ore of lead), and many other oxides, sulfides, and halides. The structure is shown in Figure 1.45; it is based on an FCC lattice of chloride anions, represented by the large open circles, with sodium cations, represented by the smaller filled circles, in all the octahedral holes. If you look carefully at the structure and compare it to the diagram of the FCC structure in Figure 1.41b, you may be able to see that there are layers of Na^+ ions between the layers of Cl^- ions so that the structure can be also be described as having alternating close-packed Cl^- and Na^+ layers. The cubic arrangement of the Na^+ and Cl^- ions in the NaCl structure is reflected in the cubic geometry of NaCl crystals.

The structure of calcium fluoride, CaF_2, which is most famous as the mineral Blue John, is based on an FCC arrangement of Ca^{2+} ions (filled circles) with F^- (open circles) in all the tetrahedral holes (Figure 1.46). Whereas each F^- ion is coordinated to four Ca^{2+} ions at the corners of the tetrahedral hole, each Ca^{2+} ion is surrounded by eight F^- ions arranged around the Ca^{2+} in the form of a cube. This structure is known as the **fluorite structure**, and is also adopted by ThO_2, CeO_2, and ZrO_2. CaF_2 itself is commercially very important as it is used in the manufacture of HF, which in turn is used to extract aluminium metal from its ore.

Diamond is an example of a covalently bonded structure, in which all the atoms are joined in an extended network. Its structure is based on an FCC lattice of carbon atoms with four more carbon atoms in half of the tetrahedral holes (Figure 1.47). Compare Figures 1.46 and 1.47 to see how carbon atoms fill alternate tetrahedral holes.

A very important mineral that has a structure based on hexagonal close packing is that of haematite, Fe_2O_3, which is the principal iron mineral used in steel making

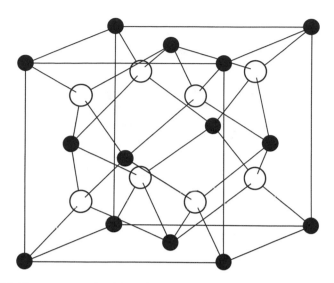

FIGURE 1.46 The structure of calcium fluoride.

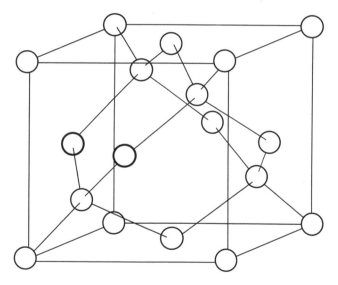

FIGURE 1.47 The structure of diamond.

(Figure 1.48). The oxide ions, O^{2-}, adopt a slightly distorted HCP arrangement (O^{2-} ions in each close-packed layer do not lie exactly in the same plane, but some are slightly raised whereas others are slightly lowered). There are 2/3 the number of Fe^{3+} ions compared to O^{2-} ions in order to balance the charge, and therefore the Fe^{3+} ions occupy only 2/3 of the octahedral holes, such that two holes out of every three in each layer are occupied. The Fe^{3+} ions (open circles) are surrounded by six O^{2-} ions (filled circles) in an octahedral arrangement, whereas the O^{2-} ions are surrounded by four Fe^{3+} ions in an approximately tetrahedral arrangement. The sites

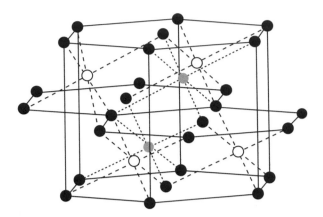

FIGURE 1.48 Structure of haematite, Fe_2O_3.

where Fe^{3+} is missing are marked with grey circles. This structure is also adopted by corundum, Al_2O_3, which is an important mineral of aluminium.

One of the most important minerals in sedimentary rocks (which you will learn about in Chapter 2, Section 2.4) is calcite. This is the more common **polymorph** (structural form) of calcium carbonate, $CaCO_3$, the other being aragonite. The structure of calcite is shown in Figure 1.49. The Ca^{2+} cations are the filled circles, and the oxygen atoms of the CO_3^{2-} anions are the open circles. For simplicity, the carbon atoms are not shown; they are found at the centre of each triangle of oxygens. The Ca^{2+} ions adopt a distorted FCC structure, and the CO_3^{2-} ions sit in the octahedral holes. This results in alternating layers of CO_3^{2-} anions and Ca^{2+} cations, with the carbonate ions all lying parallel to the plane of the layers. The structure is held together by the electrostatic attractions between the Ca^{2+} and CO_3^{2-} ions. Other carbonates such as magnesite, $MgCO_3$; dolomite, $CaMg(CO_3)_2$; siderite, $FeCO_3$; and rhodocrosite, $MnCO_3$, have similar structures to calcite.

Aragonite is a polymorph of calcium carbonate which is stable at high pressure. The structure is based on a distorted HCP arrangement of Ca^{2+} ions, again with the CO_3^{2-} ions in the octahedral holes. Over time, aragonite converts to the more stable calcite polymorph. Strontianite, $SrCO_3$, has a structure similar to aragonite.

Calcium sulfate is another evaporite mineral. It occurs naturally in both a hydrated form called gypsum, $CaSO_4.2H_2O$, which is the most common of all the sulfate minerals, and an anhydrous form called anhydrite, $CaSO_4$. The structure of gypsum is shown in Figure 1.50. It has a layer structure, with the layers being formed by Ca^{2+} and SO_4^{2-} ions. Between each layer there are water molecules. The large open circles are Ca^{2+} ions, the small open circles are oxygen atoms, the small filled circles are sulfur atoms, and the grey-shaded circles are H_2O molecules. The sulfate anion, SO_4^{2-}, has a tetrahedral arrangement of the four oxygen atoms around the sulfur atom.

The Ca^{2+} and SO_4^{2-} layers are held together strongly by ionic bonding. The bonding between the layers is weak because the bonds holding water molecules

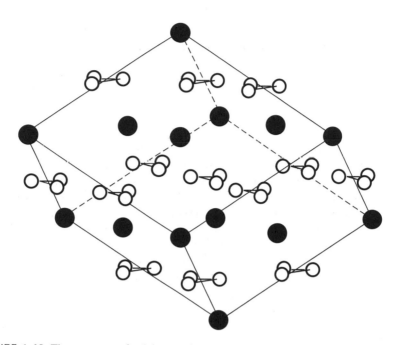

FIGURE 1.49 The structure of calcium carbonate.

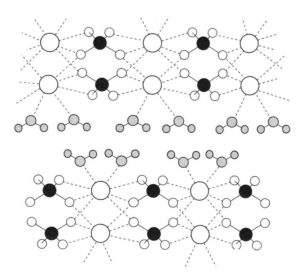

FIGURE 1.50 The structure of gypsum, $CaSO_4.2H_2O$.

together are much weaker than the electrostatic bonds holding the ions together, and therefore the layers can be easily split apart.

1.6 CHEMICAL REACTIONS AND EQUILIBRIA

Now that we have discussed how atoms and ions combine to form molecules, we need to explore how atoms, ions, and molecules react with one another in chemical reactions.

1.6.1 CHEMICAL REACTIONS

Chemical reactions occur when chemical bonds are broken and new ones made, so that the atoms or ions that were bonded together in one set of molecules are recombined in a different arrangement to form new molecules. The first important point to make is that chemical reactions must be balanced. We cannot create or destroy elements in ordinary chemical reactions (this can only happen in nuclear reactions such as the radioactivity sequences you saw in Section 1.1.5), and therefore we must have the same number of atoms or ions of each element at the end of a reaction as at the beginning. This is known as the law of conservation of matter. If there is 25 g of matter that includes 0.13 mol of carbon at the beginning of the reaction, there must be 25 g of matter that includes 0.13 mol of carbon at the end of the reaction. The matter may have changed form during the reaction, for example from a solid to a gas, but there must still be 25 g of matter, including 0.13 mol of carbon.

To illustrate the principles of chemical reactions, we will consider two examples. The first is the weathering of limestone rock, $CaCO_3$, by water, H_2O, containing dissolved carbon dioxide, CO_2. The reaction is shown in the form of a **chemical equation**, Equation 1.27. The **reactants** are on the left-hand side of the arrow, and the **products** are on the right-hand side. The physical form of each reactant is given in brackets: (s) for solid, (l) for liquid, (g) for gas, and (aq) for aqueous (i.e., dissolved in water). The **stoichiometry** of the reaction is the relative number of moles of each of the reactants used and products formed in the reaction. In this example, one mole of $CaCO_3$ reacts with one mole of CO_2 and one mole of H_2O to give one mole of Ca^{2+} ions and two moles of HCO_3^- ions. The reaction as written in Equation 1.27 is balanced because we have one calcium, two carbons, six oxygens, and two hydrogens on each side of the equation.

$$CaCO_{3(s)} + CO_{2(aq)} + H_2O_{(l)} \rightleftharpoons Ca^{2+}_{(aq)} + 2HCO^-_{3(aq)} \qquad (1.27)$$

A second example is shown in Equation 1.28, which is the combustion (burning) of methane (natural gas), CH_4, with oxygen, O_2, to give carbon dioxide and water. Equation 1.28 shows the products and reactants, but it is not balanced because there are two oxygens on the left and three on the right, whereas there are four hydrogens on the left and three on the right. To balance Equation 1.28, we need two H_2O to account for the four hydrogens in methane, and two O_2 to account for all the oxygen in CO_2 and H_2O. Equation 1.29 shows the balanced equation.

$$CH_4 + O_2 \longrightarrow CO_2 + H_2O \qquad (1.28)$$

$$CH_4 + 2O_2 \longrightarrow CO_2 + 2H_2O \qquad (1.29)$$

SELF-ASSESSMENT QUESTION

Q1.26 Balance the equation for the combustion of butane (bottled gas), C_4H_{10}, with oxygen, which gives carbon dioxide and water. Hint: Start by calculating how many moles of CO_2 and H_2O are formed for every mole of C_4H_{10}, and then decide how many moles of O_2 are required.

1.6.2 THE ENERGY OF CHEMICAL REACTIONS

Simply writing down a chemical equation on the page is no guarantee that the reactants will actually react with each other. There are two reasons why this may be so. Firstly, more energy may have to be put in to break the chemical bonds in the reactants than is released when the new bonds are made to form the products. In chemical reactions, some chemical bonds are broken and new ones are made. A reaction is energetically favourable when more energy is given out on forming the new bonds than has to be put in to break the original bonds. Such a reaction is called an **exothermic reaction**, and most reactions that take place readily are exothermic reactions. A reaction in which more energy has to be put in to break bonds than is released when the new bonds are made is called an **endothermic reaction**. Some endothermic reactions do take place, usually if one of the products is a gas, but the reasons for this are beyond the scope of this book. The second reason that a reaction might not take place, or might be very slow, is that although the reaction may be exothermic, the amount of energy required actually to break the chemical bonds in the reactants is too high: the **activation energy** (i.e., the energy hurdle) may be too high for the reaction to start.

As an example, consider the combustion of methane in air to give carbon dioxide and water, shown already in Equation 1.29. The structures of the reactants and products can be drawn out to show the bonds in each molecule (Figure 1.51). The bond energies of the different types of bond are given in Table 1.8, but we also have to count up the number of bonds of each type that are made or broken. In the combustion of methane, four C-H bonds (414 kJ mol^{-1}) and two O=O bonds (498 kJ mol^{-1}) will be broken, and two C=O bonds (803 kJ mol^{-1}) and four H-O bonds (464 kJ mol^{-1}) will be made. Using this information, the energy, which is properly

FIGURE 1.51 Combustion of methane showing the bonds in the reactants and products.

called the **enthalpy** or **heat energy**, needed to break the bonds in methane and in oxygen can be calculated as follows:

Energy required to break bonds = $(4 \times 414) + (2 \times 498)$ kJ mol^{-1} = 2652 kJ mol^{-1}

The energy given out on making the bonds in carbon dioxide and water can be calculated as follows:

Energy given out on making bonds = $(2 \times 803) + (4 \times 464)$ kJ mol^{-1} = 3462 kJ mol^{-1}

The change in energy, which is properly called the **enthalpy change**, for this reaction is the energy put in to break the original bonds minus the energy given out on making new bonds. So for this reaction:

Enthalpy change = 2652 − 3462 kJ mol^{-1} = −810 kJ mol^{-1}

If the enthalpy change has a negative sign, this means that overall heat energy is given out during a chemical reaction; therefore, the reaction is an exothermic reaction, and it will usually proceed in the forward direction. If the enthalpy change has a positive sign, it means that overall heat energy is put in during a chemical reaction; the reaction is an endothermic reaction, and it will usually not proceed in the forward direction. For the combustion of methane, there is a negative enthalpy change and, therefore, heat energy is given out and the reaction proceeds in the forward direction. We know this intuitively, of course, because methane is burned in power stations to produce electricity, and in homes for cooking and central heating. The total energy or **free energy** of the reaction includes another term called **entropy**, but we will not consider it here; it will not affect the total energy much in this system.

SELF-ASSESSMENT QUESTION

Q1.27 Petrol (gasoline) contains a mix of compounds, many of which are structural isomers of octane, C_8H_{18}.
  (i)   Draw the structural formula of octane, which is a straight-chain alkane, and use this to determine how many C-C and C-H bonds are broken when petrol is burned.
  (ii) The reaction for the combustion of octane is given as follows:

$$C_8H_{18} + \tfrac{25}{2}O_2 \longrightarrow 8CO_2 + 9H_2O$$

  Determine the number of O=O bonds that are broken, and the number of C=O and H-O bonds that are made when petrol is burned.
  (iii) Use the information in Table 1.8 to calculate the enthalpy change on burning octane with oxygen to give carbon dioxide, CO_2, and water, H_2O.

1.6.3 CHEMICAL EQUILIBRIA AND LE CHATELIER'S PRINCIPLE

The combustion of methane shown in Equation 1.29 is a reaction that can be considered as going to completion; i.e., as long as there is some oxygen and methane present, the reaction will continue. On the other hand, if the products of the reaction, carbon dioxide and water, are mixed, they will not recombine to form butane and oxygen. This combustion reaction is essentially a one-way system. It is an exothermic reaction, and so the reverse reaction is an endothermic reaction, because very much more energy would have to be put in to break C=O and H-O bonds than would be got back at the end of the reaction in forming C-C and C-H bonds. In principle, if enough energy were put into a system containing CO_2 and water in the appropriate proportions, butane (or its isomers) and oxygen could be formed, but in practice this does not occur. (Carbon dioxide and water will react together to form cellulose, $(CH_2O)_n$, but under the very specific conditions of photosynthesis in plants.) By contrast, the weathering of limestone (Equation 1.27) is a readily reversible reaction: calcium carbonate will precipitate out of a solution containing Ca^{2+} ions and HCO_3^- ions given the right conditions (e.g., water dripping from the roof of a limestone cave forming stalactites and stalagmites, or warm shallow tropical waters, for example).

Reversible reactions in a closed system (that is, one in which none of the reactants or products can escape to the surroundings) will eventually come to an equilibrium position. Initially, when the reactants are mixed, the reaction proceeds in the forward direction until the conditions for the reverse reaction become favourable. The forward and reverse reactions then proceed at the same rate so that the overall concentrations of the reactants and products do not change. The reaction is then said to be in **equilibrium**. The reaction will stay in equilibrium until the system is perturbed, either by a change in the temperature of the system or by adding or removing one of the reagents, i.e., by changing the concentrations of one (or more) of the reagents. The position of the equilibrium in a reversible reaction is controlled by the **equilibrium constant**, K, which relates the concentrations of the reactants to the concentrations of the products. Consider the following reaction, in which m molecules of substance A react with n molecules of substance B to give p molecules of X and q molecules of Y:

$$mA + nB \rightarrow pX + qY$$

The equilibrium constant for this reaction is given in Equation 1.30. The notation "[A]" means "the concentration of A"; and similarly [B] means "the concentration of B," etc.

$$K = \frac{[X]^p \times [Y]^q}{[A]^m \times [B]^n} \qquad (1.30)$$

Using Equation 1.30 we can see that if either [A] or [B] is increased by adding more reactant to the reaction, the value of ([A]^m × [B]^n) will increase, and therefore the value of $[X]^p \times [Y]^q/([A]^m \times [B]^n)$ will be reduced. The reaction therefore will proceed in the forward direction so that [X] and [Y] increase, until equilibrium is re-established, i.e., until the value of $[X]^p \times [Y]^q/([A]^m \times [B]^n)$ is once more equal to K. The same effect will occur if either [X] or [Y] is reduced by removing some product. On the other hand, if either [X] or [Y] is increased by adding more product, the value of $[X]^p \times [Y]^q/([A]^m \times [B]^n)$ will also be increased. The reaction will proceed in the reverse direction so that [A] and [B] increase until the value of $[X]^p \times [Y]^q/([A]^m \times [B]^n)$ is once more equal to K, and equilibrium is reestablished. The same effect will occur if either [A] or [B] is reduced by removing some of the reactant.

We can illustrate the preceding discussion by reference to an example. In the oceans, the equilibrium between carbonate ions in solution, CO_3^{2-}, and precipitated $CaCO_3$ (Equation 1.31) is important for organisms such as corals that have calcite skeletons. If the concentration of CO_3^{2-} ions decreases, then precipitated calcium $CaCO_3$ will dissolve to increase the concentration of CO_3^{2-} ions again. This is likely to be one effect of increasing atmospheric CO_2 levels due to global warming, as you will see in Chapter 3, Section 3.3.3.

$$Ca^{2+}_{(aq)} + CO^{2-}_{3(aq)} \rightleftharpoons CaCO_{3(s)} \qquad (1.31)$$

K for the reaction at 25 °C (298 K) is 2.6×10^8, and the expression for K is given in Equation 1.32. The concentration of Ca^{2+} in the oceans is about 407 mg kg^{-1} (Table 3.4), which is 0.0102 mol L^{-1}, whereas the value of $[CaCO_3]$ is set to 1 because $CaCO_3$ is a solid. (You do not need to know why this is so.) We can therefore calculate the concentration of carbonate ions in sea water, as shown in Equation 1.33. You will learn more about the carbonate system in Chapter 3, Section 3.3.3.

$$K = \frac{[CaCO_3]}{[Ca^{2+}] \times [CO_3^{2-}]} \qquad (1.32)$$

$$[CO_3^{2-}] = \frac{[CaCO_3]}{[Ca^{2+}] \times K} = \frac{1}{0.0102 \times 2.6 \times 10^8} \text{ mol } L^{-1} = 3.8 \times 10^{-7} \text{ mol } L^{-1} \qquad (1.33)$$

SELF-ASSESSMENT QUESTION

Q1.28   (i)   Write down the expression for K for the following reaction:

$$2SO_2 \text{ (g)} + O_2 \text{ (g)} \leftrightarrow 2SO_3 \text{ (g)}$$

       (ii) SO_2 and O_2 were reacted together in a sealed 1 L flask at 1000 K (727°C). At the end of the reaction there were 0.063 moles of SO_2, 0.721 moles of O_2, and 0.895 moles of SO_3. Calculate K for this reaction.

The way that an equilibrium reaction responds to a decrease in the concentration of one of the reagents by reacting so as to increase the concentration of that reagent, or responds to an increase in the concentration of a reagent by reacting to decrease the concentration of that reagent, is an example of **Le Chatelier's Principle**. This principle states that if an equilibrium reaction is perturbed in any way (by adding or removing reagents, changing the temperature, or adding a catalyst), then the reaction will proceed in the direction that minimises the change until equilibrium is re-established.

The dissolution of limestone rock is actually an example of a reversible reaction that does not achieve equilibrium. This is because as $CaCO_3$ is dissolved, the water soaks away, thus removing both products (Ca^{2+} and HCO_3^-). Therefore, the reaction continues in the forward direction to dissolve more $CaCO_3$. The result is the formation of extensive networks of underground caverns and channels in areas of limestone geology.

1.7 SUMMARY

In this chapter you should have learned that:

Atoms are made from three fundamental particles: protons and neutrons in the central nucleus, and electrons in orbitals around the nucleus. Protons have a single positive charge, and electrons have a single negative charge.

An element is defined by the number of protons in its nucleus. Isotopes of an element have the same number of protons but different numbers of neutrons.

Radioactive isotopes decay to other elements by various processes that change the number of protons, neutrons, and electrons in the isotope.

Matter can be in the form of a plasma, gas, liquid, or solid. The state that a particular substance is in depends on pressure and temperature, and this can be represented on a phase diagram. Transitions from one state of matter to another are called phase transitions.

Pure substances cannot be split into simpler components by physical separation, whereas mixtures can be split into simpler substances by physical separation. Compounds are substances in which two or more atoms of different elements are chemically combined.

A molecule is a substance in which two or more atoms are linked by a chemical bond. Atoms can form a chemical bond by sharing electrons. This type of bond is called a covalent bond.

In some covalent bonds the electrons are not equally shared between the two atoms because one of the atoms is more electronegative than the other. Such bonds are called polar bonds, and the most important type of polar bond is the hydrogen bond.

Atoms or molecules can gain or lose electrons to form ions. Ions of opposite charge form ionic bonds by simple electrostatic attraction.

In metals, the atoms pack together in a sea of electrons, and the atoms are held together by their mutual attraction for the electrons.

Carbon can form over two million different compounds, because of its unique ability to form chemical bonds with itself and other elements. These compounds (with a very few exceptions) are called organic compounds, and they are classified according to different functional groups in the molecules.

Inorganic compounds are compounds of all elements except carbon, plus a very few carbon-containing compounds.

Compounds can be made up of discrete molecules or of extended networks of atoms or ions.

Chemical reactions occur when chemical bonds are broken and made, but the number of atoms or ions of each element must be the same at the end of a chemical reaction as at the beginning.

Chemical equations show the reactants and products involved in a reaction. Not all chemical reactions will proceed in the forward direction. Most that do are exothermic, i.e., they give out heat. Endothermic reactions take in heat and mostly do not proceed in the forward direction.

Some chemical reactions are reversible and come to equilibrium. If the equilibrium is disturbed by adding or removing species, the reaction adjusts according to Le Chatelier's principle.

Now try the following questions to test your understanding of the material covered in this chapter.

SELF-ASSESSMENT QUESTIONS

Q1.29 Element M has two isotopes (among others) that are given as $^{32}_{16}M$ and $^{33}_{16}M$.
  (i)   Identify element M
  (ii) Give the number of protons, neutrons, and electrons in each of these two isotopes.
Q1.30 (i)   What ions are present in dolomite, $CaMg(CO_3)_2$?
  (ii) Using the atomic weights given in the periodic table rounded up to two decimal places, calculate the RMM of dolomite.
  (iii) Express the RMM of dolomite as a molar mass.
  (iv) Calculate the number of moles of dolomite in 2.5 kg of sample.
Q1.31 (i)   What ions are present in sodium sulfate, Na_2SO_4?
  (ii) Calculate the concentration of a solution of that has 10.3 g of sodium sulfate dissolved in 250 mL of water.
  (iii) What is the concentration of Na^+ ions in this solution?
  (iv) Calculate the concentration of Na^+ in units of mg L^{-1}.
Q1.32 Identify the functional groups that are present in the following organic molecules, and therefore what type of compound they are.

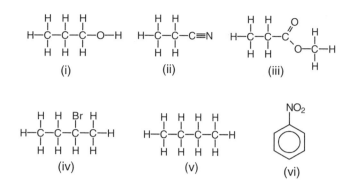

FIGURE 1.52 Organic structures for SAQ1.32.

Q 1.33 An important side reaction when fossil fuel is burned at high temperature is the formation of NO by direct combination of O_2 and N_2.

$$N_2 + O_2 \rightarrow 2NO$$

(i) Write down the expression for K, the equilibrium constant for this reaction in terms of the concentrations of the species involved in the reaction.

(ii) In reactions involving gases, concentrations are given in terms of partial pressures, i.e., the fraction of atmospheric pressure that is due to each gas. The partial pressure of N_2, p_{N2}, is 0.78 atm, whereas the partial pressure of O_2, p_{O2}, is 0.21 atm. The partial pressure of NO is given by p_{NO}. Rewrite your answer to (i) in terms of partial pressures.

(iii) The value of K for this reaction at 1500 K (1227°C) is 1.0×10^{-5}. Assuming that the partial pressures of N_2 and O_2 change very little, calculate the partial pressure of NO formed in this reaction.

2 Earth

In this chapter we will be considering the structures of silicate and aluminosilicate minerals, which form the bulk of the Earth's outer layers, and the types of rocks that are formed from these and other minerals. We will also see how these rocks are broken down and recycled. We will discuss the nature of soil, and consider aspects of soil pollution. First, however, we will learn about the formation and structure of the Earth.

2.1 FORMATION OF THE EARTH

The generally accepted model for the formation of the Sun and the solar system is called the **Standard Model**. About 5000 million years ago, a dense, rotating cloud of interstellar gas and dust collapsed under its own gravity to form the Sun, which was surrounded by a disc of residual material called the **solar nebula**. The Earth formed some 4550 million years ago from the disc of dust and gas surrounding the newly formed Sun. As the solar nebula cooled to below about 1800 K, solid material began to condense out as tiny grains of dust. Below about 1400 K, metallic iron and silicate minerals condensed, to be followed, on further cooling, by more volatile elements and compounds. Gradually, these grains coalesced to form larger and larger rocks. Within about 10,000 years, planetesimals of about 10 km radius, with masses about 10^{15} kg, had formed. Some of these planetesimals began to grow faster than others and, within about 100,000 years, had accumulated sufficient material to form planetary embryos, about the size of the Moon or Mercury, and with masses of about 10^{22} to 10^{23} kg. Eventually, most of the remaining planetary embryos had collided and coalesced to form the planets we know today. (Pluto is unexpectedly small, however, and has now been reclassified as a **dwarf planet**.)

Heat generated in the embryo Earth by bombardment and by short-lived radioactive isotopes would have caused melting of much, perhaps even all, of the Earth. As a result of this melting, denser material — principally, metallic iron with some nickel — would have been able to sink to the centre of the planet, forming the Earth's core. The Moon formed very shortly after the formation of the Earth. It is now widely believed that the Moon was formed when a Mars-sized object hit the Earth, sending huge quantities of material into orbit around the Earth, which rapidly coalesced to form the Moon.

Volcanic acitvity was extensive on the early Earth because of the hot rocks close to the Earth's surface, and this spewed carbon dioxide, CO_2, water, H_2O, and hydrogen chloride, HCl into the atmosphere. Liquid water is now believed to have been present from an early point in the Earth's history, formed by the condensation of large amounts of water vapour. The Earth and other planets continued to suffer from frequent impacts until about 3900 million years ago, and this bombardment would have brought huge quantities of water to Earth in the form of icy comets. It is currently a matter of debate as to how much of the water in the Earth's oceans came from terrestrial volcanic activity, and how much came from comets.

2.1.1 THE STRUCTURE OF THE EARTH

The Earth is shaped like a very slightly squashed sphere: it is an oblate spheroid, having a slightly larger diameter at the equator (12,756 km) than at the poles (12,714 km). The Earth has a layered structure rather like an onion, which is shown in Figure 2.1. We of, course, cannot see into the Earth to find out its structure and what it is made from. Most evidence for the structure of the Earth has come from studying earthquakes and the way that seismic waves produced in earthquakes (or the occasional nuclear explosion) pass through the Earth. This has been combined with other information, such as mineral densities and the elemental compositions of meteorites, to build up the picture as we know it today.

At the centre of the Earth is the core, which has a radius of 3480 km and which is mainly metallic iron, with a small percentage of nickel, and perhaps 5–15% of lighter elements (possibly oxygen, sulfur, or potassium). The inner core is solid, and has a radius of about 1220 km. The outer core is molten and is about 2260 km thick. Over time, the outer core will decrease in size as the inner core continues to crystallise. It is the motion of liquid iron in the outer core that generates the Earth's magnetic field and proves that molten iron must be present.

The mantle lies above the core and is about 2890 km thick. It is divided into the upper mantle, which is about 670 km thick, and the lower mantle. The mantle

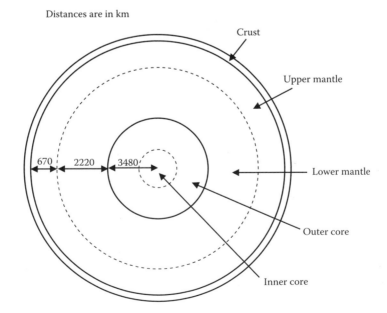

FIGURE 2.1 The structure of the Earth (not to scale; distances are in km).

TABLE 2.1
Composition of the Earth's Crust and Mantle

Element	Symbol	Crustal wt %	Mantle wt %
Oxygen	Si	46.6	44.0
Silicon	O	27.7	21.1
Magnesium	Mg	2.1	23.4
Iron	Fe	5.0	7.1
Aluminium	Al	8.1	1.7
Sodium	Na	2.8	0.5
Calcium	Ca	3.6	2.0
Potassium	K	2.6	
Others		1.4	0.2

is solid, but is nevertheless able to deform and therefore convect heat as hot material rises towards the surface of the Earth. The elemental composition of the mantle (by weight percent) is given in Table 2.1, and you can see that the most abundant element is oxygen, O, followed by magnesium, Mg, and silicon, Si. Together they account for nearly 89% of the mass of the mantle. The upper mantle is formed from a rock called **peridotite**, which mainly consists of the mineral **olivine**. The lower mantle is believed to be formed from minerals with the **perovskite** structure, mainly magnesium silicate perovskite ($MgSiO_3$) and calcium silicate perovskite ($CaSiO_3$). You will learn more about olivine and perovskite in Section 2.2. Our knowledge of the composition of the mantle comes in part from **ophiolites**, which are sections of the upper mantle and **crust** that have been pushed up onto the surface of the Earth.

Above the mantle lies the Earth's outer layer, called the crust, which, in contrast to the core and mantle, is very thin. Oceanic crust, which lies beneath the world's oceans, is pretty uniform in thickness, on average only about 7 km. Continental crust, which forms the Earth's land masses, is more variable in thickness, ranging from 10 to 90 km, with an average thickness of about 35 km. The crust is mainly composed of silicate rocks such as basalts and granites, which are composed of minerals such as feldspars, micas, and quartz. You will learn more about these minerals in Section 2.2. The elemental composition of the crust is given in Table 2.1, and you can see that, compared to the mantle, there is very much less magnesium, a little more silicon, and proportionately much more aluminium, Al.

The other inner planets of the solar system (Mercury, Venus, and Mars) share with the Earth the basic structure of a metallic core with a thick outer layer of largely silicate rock, and hence they are collectively referred to as the terrestrial planets.

Box 2.1 Dating The Earth

You learnt in Chapter 1, Section 1.6, how radioactivity can be employed to date rocks and archaeological artefacts, using different combinations of radioisotopes depending on the age of the rocks. However, how is it possible to determine the age of the Earth when there are no rocks left (or that have been discovered) from the Earth's surface when it first formed? The oldest intact rocks discovered to date are the Acasta Gneiss from northwestern Canada, which are 4030 million years old. The oldest known minerals are zircon crystals from Jack Hills, western Australia, which have survived in younger rocks and are 4400 millon years old. How do we know how much older the Earth is than these rocks and minerals?

The answer comes, perhaps surprisingly, from space. Between the orbits of Earth and Mars lies the asteroid belt, which is made up of planetesimals together with fragments formed as a result of collisions between planetesimals. The material in the asteroid belt never coalesced due to the gravitational influence of Jupiter. Occasionally, collisions in the asteroid belt cause rocks to be ejected from the belt, and some of these subsequently fall to Earth as meteorites.

Meteorites fall into three basic categories: iron, stony iron, and stony. Stony-iron meteorites in turn are subdivided into chondrites, or chondritic meteorites, and achondrites, or achondritic meteorites. It can be shown that iron, stony iron, and achondrites must have formed *after* their parent bodies in the early solar system had differentiated, i.e., after they had formed an iron core surrounded by a rocky mantle. By contrast, chondrites must have formed *before* their parent bodies differentiated, as the rock matrix still contains tiny grains or chondrites of iron–nickel alloy.

The most primitive type of chondrites are believed to be a type called carbonaceous chondrites. These have the highest levels of carbon and water, and would have come from planetesimals that had not yet melted. Carbonaceous chondrites are believed to have the same composition as the early Earth and to be formed of the same material from which the Earth condensed. These meteorites have been dated radiometrically (using their Rb/Sr isotope ratios), and the ages come out very consistently at 4550 million years.

2.2 THE STRUCTURES OF SILICATE MINERALS

The mantle and crust of the Earth are largely composed of silicate minerals. A mineral is a naturally occurring crystalline material that has a specific or limited range of chemical compositions, and silicate minerals are minerals based on the elements silicon and oxygen. Oxygen is the most abundant, and silicon is the second most abundant, element in the Earth's crust, and together account for 74.3% by mass, as shown by the data in Table 2.1.

Most silicate minerals contain, as their main building block, the silicate group, $[SiO_4]^{4-}$. This consists of a silicon atom bonded to four oxygen atoms that surround the central silicon in the shape of a tetrahedron, as shown in Figure 2.2a. The −4 charge on the silicate group arises from considering every oxygen atom to carry a charge of −2, and every silicon atom, to carry a charge of +4, but in fact, the bonding between Si and O is largely covalent, although rather polarised. The $[SiO_4]^{4-}$ tetrahedron can be considered as being derived from silicic acid, H_4SiO_4 in which hydrogen ions (H^+) are balancing the charge on the $[SiO_4]^{4-}$ ion, but in minerals the electrical charge will be balanced with metal cations.

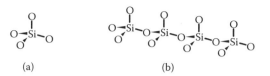

(a)   (b)

FIGURE 2.2 Silicate tetrahedra: (a) an isolated $[SiO_4]^{4-}$ tetrahedron; (b) tetrahedra linked through corners to form a chain.

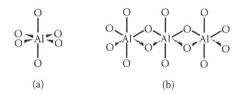

(a)   (b)

FIGURE 2.3 Aluminate octahedra: (a) an isolated $[AlO_4(OH)_2]$ octahedron; (b) octahedra linked through two corners to form a sheet.

Silicate tetrahedra can be linked by sharing oxygen atoms, as shown in Figure 2.2b. Each oxygen atom can be shared between two tetrahedra so that any one $[SiO_4]^{4-}$ tetrahedron may be joined to up to four other tetrahedra. Tetrahedra can be linked in many different arrangements forming chains, sheets, and three-dimensional networks, as you will see in the following sections. We will be considering each type of arrangement in turn. This large variety of arrangements gives rise to a very large variety of minerals.

The ratio of silicon to oxygen atoms in a mineral consisting of isolated $[SiO_4]^{4-}$ tetrahedra is 1:4. The silicon-to-oxygen ratio in a mineral increases as the degree of linking increases, to a maximum of 1:2, in a mineral in which every tetrahedron is linked to four others in a three-dimensional network, such as in the mineral quartz, SiO_2.

Some silicate minerals such as clays have a second building block in the form of either $[AlO_3(OH)_3]^{6-}$ or $[MgO_2(OH)_4]^{6-}$ groups in which the central metal atom (either Al^{3+} or Mg^{2+}) is surrounded by six oxygen atoms in the shape of an octahedron (Figure 2.3a). These octahedra are typically linked together to form sheets by having an octahedron share two adjacent oxygen atoms with another octahedron so that the octahedra are joined along an edge (Figure 2.3b). You will learn more about the structures of clays and related minerals in Section 2.2.4.

2.2.1 SILICATES FORMED FROM ISOLATED TETRAHEDRA — ORTHOSILICATES

The Earth's upper mantle is formed of a rock called peridotite, which largely consists of the mineral olivine. Olivine is a very important example of a silicate formed from isolated $[SiO_4]^{4-}$ tetrahedra. The electrical charge on $[SiO_4]^{4-}$ is balanced with Mg^{2+} and Fe^{2+} ions, giving olivine a chemical formula of $(Mg,Fe)_2SiO_4$. The brackets indicate that the total number of Mg^{2+} plus Fe^{2+} ions in the formula is two, but that

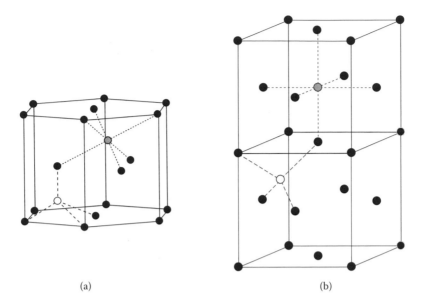

(a)   (b)

FIGURE 2.4 Structure of orthosilicates: (a) structure of olivine showing oxygens (filled circles) in an HCP arrangement with iron or magnesium (grey circle) in an octahedral hole and silicon (open circle) in a tetrahedral hole; (b) structure of spinel showing oxygens (filled circles) in an FCC arrangement with iron or magnesium (grey circle) in a tetrahedral hole and silicon (open circle) in a tetrahedral hole.

Mg^{2+} and Fe^{2+} can occur in any proportion. Magnesium and iron form what is called a **solid solution**; that is, the composition of olivine can vary from pure Mg_2SiO_4 (forsterite) to pure Fe_2SiO_4 (fayalite) in any proportion of Mg_2SiO_4 to Fe_2SiO_4. Magnesium and iron are able to do this because not only do they carry the same +2 charge but their radii are very similar (Mg^{2+} has a radius of 86 pm, and Fe^{2+}, a radius of 92 pm), so it is easy for Fe^{2+} to substitute for Mg^{2+} and vice versa without the solid structure being distorted. You will learn more about the forsterite–fayalite solid solution series in Section 2.3.3.

The structure of olivine is shown in Figure 2.4a. The $[SiO_4]^{4-}$ tetrahedra pack together in such a way that the oxygen atoms form a hexagonal close-packed arrangement. The silicon atoms occupy one-fourth of the tetrahedral holes so that each silicon is surrounded by four oxygens, and the Mg^{2+} or Fe^{2+} ions occupy half of the octahedral holes, so that each Mg^{2+} or Fe^{2+} ion is surrounded by six O atoms. (See Chapter 1, Section 1.4.2, for a description of hexagonal close packing, octahedral and tetrahedral holes.)

At pressures and temperatures equal to a depth of about 400 km, olivine becomes unstable and converts to the mineral spinel, which is slightly denser than olivine. This mineral is also formed from isolated $[SiO_4]^{4-}$ tetrahedra, and its structure is shown in Figure 2.4b. In the solid state these tetrahedra pack together in such a way that the oxygen atoms form a close-packed or face-centred cubic arrangement. Again, silicon atoms occupy one-fourth of the tetrahedral holes, and Mg^{2+} or Fe^{2+} ions occupy half of the octahedral holes. (See Section 1.4.2 for a description of cubic close packing.)

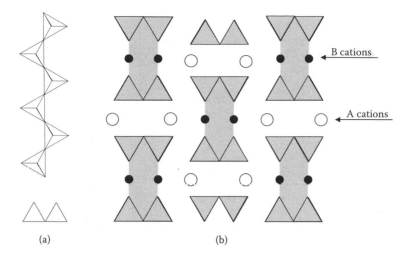

FIGURE 2.5 Structure of pyroxenes: (a) chain of tetrahedra, with (below) end-on view; (b) chains held together by B cations to form I-beams (highlighted in grey) and I-beams held together by A cations.

2.2.2 SINGLE-CHAIN SILICATES — PYROXENES

$[SiO_4]^{4-}$ tetrahedra can join to form single chains by sharing two of their four oxygen atoms, as shown in Figure 2.5a. These chains form the basis of the **pyroxene** family of minerals, which are an important component of basalt, a rock that is a major constituent of the Earth's crust. In the solid state, pairs of chains are arranged with the apical oxygen atoms (i.e., the tops of the $[SiO_4]^{4-}$ tetrahedra) pointing inwards, and they are held together by mutual electrostatic attraction to M^{2+} ions. These paired chain structures are called I-beams because of their end-on appearance. The I-beams in turn are held together by more M^{2+} ions, as shown in Figure 2.5b. There are, thus, two types of positions for the M^{2+} ions: the A sites, which hold the I-beams together, and the B sites, which hold the chains together in the I-beams. Pyroxenes have the general formula $ABSi_2O_6$, and the chains themselves have the formula $[SiO_3]^{2-}$.

Two important pyroxenes are **clinopyroxene** and **orthopyroxene**. Clinopyroxenes have the idealised general formula $Ca(Mg,Fe)Si_2O_6$. Complete solid solution between Mg and Fe is possible so that any particular sample could have a composition ranging from $CaMgSi_2O_6$ (diopside) to $CaFeSi_2O_6$ (hedenbergite). The Ca^{2+} ions are found in the A sites, whereas the smaller Mg^{2+} and Fe^{2+} ions are found in the B sites. Orthopyroxenes have the idealised general formula $(Mg,Fe)_2Si_2O_6$, and Fe^{2+} and Mg^{2+} can each occupy both A and B sites. The two end members of this solid solution series are $MgSiO_3$, enstatite, and $FeSiO_3$, ferrosilite.

2.2.3 DOUBLE-CHAIN SILICATES — AMPHIBOLES

Tetrahedra can be linked to form double chains, as shown in Figure 2.6a. Some of these double-chain structures give rise to the class of minerals called **amphiboles**. The double chains are held together to form I-beams by small cations such as Mg^{2+},

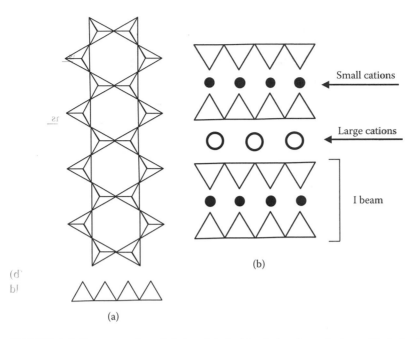

(b)

(a)

FIGURE 2.6 Structure of amphiboles: (a) double chain of tetrahedra, with (below) end-on view; (b) double chains held together by small cations to form I-beams and I-beams held together by large cations.

Fe^{2+}, or Al^{3+}, just as in pyroxenes, whereas the I-beams are held together by larger cations such as Ca^{2+}, Na$^+$, or K$^+$ (Figure 2.6b). The double chains of amphiboles have the general formula $[Si_4O_{11}]^{6-}$. An example of an amphibole is tremolite, $Ca_2Mg_5Si_8O_{22}(OH)_2$. In this mineral, the hydroxyl groups coordinate to Mg^{2+} ions in the centre of the I-beams.

An important feature of amphiboles is the variety of compositions that can occur because of ion replacement. For example, common hornblende has the formula $Ca_2Mg_4Al(Si_7Al)O_{22}(OH)_2$. In this mineral an Al^{3+} ion has replaced one Si^{4+} ion in the double chain, and to maintain the overall charge balance, an Al^{3+} ion has also replaced an Mg^{2+} ion. Another example is edenite, which has the formula $NaCa_2Mg_5(Si_7Al)O_{22}(OH)_2$. As in common hornblende, Al^{3+} has replaced Si^{4+} in the double chain, but charge balance has been maintained this time by adding an Na$^+$ ion, which sits between the chains.

2.2.4 Sheet Silicates — Micas, Clays, and Talc

Sheets of tetrahedra can be formed by linking each $[SiO_4]^{4-}$ tetrahedron to three others, as shown in Figure 2.7. These sheets have the repeat unit $[Si_4O_{10}]^{4-}$, and form the basis of a large number of minerals, including micas, clays, and talc. In these minerals, sheets of linked $[SiO_4]^{4-}$ tetrahedra are combined with sheets of linked $[AlO_3(OH)_3]^{6-}$ or $[MgO_4(OH)_2]^{8-}$ octahedra. Sheets of linked $[AlO_3(OH)_3]^{6-}$ octahedra are called **dioctahedral** sheets, and are shown in Figure 2.8a. They are formed

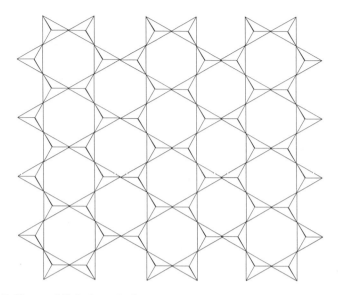

FIGURE 2.7 Sheets of linked tetrahedra.

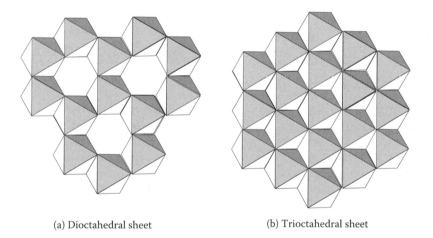

(a) Dioctahedral sheet   (b) Trioctahedral sheet

FIGURE 2.8 Sheets of linked octahedra: (a) a dioctahedral sheet with each oxygen atom part of two octahedra; (b) a trioctahedral sheet with each oxygen atom part of three octahedra.

by joining $[AlO_3(OH)_3]^{6-}$ octahedra along their edges to three other octahedra, so that each oxygen is part of two (hence di) octahedra. Sheets of linked $[MgO_4(OH)_2]^{8-}$ octahedra are called **trioctahedral** sheets, and are shown in Figure 2.8b. They are formed by joining $[MgO_4(OH)_2]^{8-}$ octahedra along their edges to six other octahedra, so that each oxygen is part of three (hence tri) octahedra. Dioctahedral sheets can also be described as layers of the mineral gibbsite, $Al(OH)_3$, and trioctahedral sheets can also be described as layers of the mineral brucite, $Mg(OH)_2$.

Octahedral (O) and tetrahedral (T) sheets are linked through the apical oxygen atoms of the silicate tetrahedra so that each apical oxygen also becomes part of an

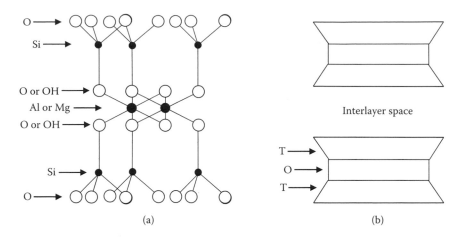

FIGURE 2.9 Sheet silicates with TOT layers: (a) showing sheets of silicate tetrahedra (above and below) linked to a dioctahedral or troctahedral sheet (middle) to give a (tetrahedral/octahedral/tetrahedral) TOT layer; (b) showing two TOT layers with the interlayer space between them.

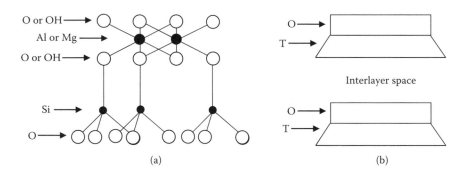

FIGURE 2.10 Sheet silicates with TO layers: (a) showing a sheet of silicate tetrahedra (below) linked to a dioctahedral or troctahedral sheet (above) to give a TO layer; (b) showing two TO layers with the interlayer space between them.

octahedron. The linking of two tetrahedral sheets and one octahedral sheet via apical oxygen atoms to form a tetrahedral/octahedral/tetrahedral (TOT) layer is shown in Figure 2.9a. These TOT layers can then stack, leaving an interlayer space between them, as shown in Figure 2.9b. The linking of one tetrahedral and one octahedral sheet to form a tetrahedral/octahedral (TO) layer is shown in Figure 2.10a, whereas Figure 2.10b shows the stacking of two TO layers with the interlayer space between them.

Substitution of Si^{4+} in a tetrahedral sheet with Al^{3+}, or substitution of Al^{3+} with M^{2+} in an octahedral sheet, will give rise to layers that have an overall negative charge. This charge must be balanced by cations, typically Ca^{2+}, Na^+, or K^+, which are located in the interlayer space between each TOT or TO layer. In some minerals,

TABLE 2.2
Structures of Sheet Aluminosilicates

Structure	Silicates with Al(OH)₃ Dioctahedral Layers	Silicates with Mg(OH)₂ Trioctahedral layers
interlayer cations → O O O	Muscovite	Biotite
no interlayer cations →	Pyrophyllite	Talc
exchangeable cations and water → O O O	Smectite (Montmorillonite)	Vermiculite
no interlayer cations →	Kaolinite	Serpentine

these cations can be exchanged for other cations (e.g., H^+), and they are then known as exchangeable cations. Water can also be found between the layers in some minerals. Some of the common sheet silicate minerals that have either TOT or TO layer arrangements are shown in Table 2.2. Vermiculite **exfoliates** if it is heated rapidly to above about 300°C; that is, the TOT layers buckle and separate as water vapour is forced out. The structure increases in thickness by as much as 30-fold along the direction perpendicular to the plane of the layers, and the resulting low-density material has very good thermal insulation properties and is also widely used as packaging material.

One consequence of the TOT or TO layer structure is that the bonding between the layers is weak, and thus the layers can be easily separated. Mica is well known for its perfect cleavage, which is a result of the weak bonding between the TOT layers. Therefore, mica can be split into very thin sheets. Talc has no interlayer cations, and the layers can easily slide over each other because there is no electrostatic attraction to hold them together; this is why talc is very soft.

2.2.5 FRAMEWORK OR TECTOSILICATES — SILICA, FELDSPARS, AND ZEOLITES

Three-dimensional framework structures can be formed by linking each $[SiO_4]^{4-}$ tetrahedron in the structure to four others. Tetrahedra linked in this way have the general formula SiO_2, which is that of silica. The most familiar silica mineral is quartz, which is the second most abundant mineral in the Earth's crust. There are in fact two forms of quartz, namely, α-quartz, which is stable up to 573°C at atmospheric pressure, and β-quartz, which is stable from 573 to 870 °C at atmospheric pressure. One way of describing the structure of quartz is that it is based on helices (shown in Figure 2.11a) with six-sided repeat units (Figure 2.11b). These helices are interlinked to form the three-dimensional network. The six-sided structure at the atomic level is reflected in the shape of quartz crystals, which are also six-sided, and the 120° internal angles between the crystal faces. The α-form is converted to the β-form by rotating the $[SiO_4]^{4-}$ tetrahedra. No chemical bonds are broken or made in this conversion, and so it takes place very quickly and requires very little energy. This kind of conversion is called a **displacive phase transition**.

Other polymorphs or structural forms of SiO_2 are stable at different temperatures and pressures, as shown in the phase diagram in Figure 2.12. Tridymite, cristobalite, and coesite also have structures based on linked tetrahedra. The very-high-pressure

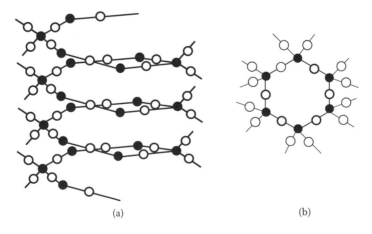

(a)   (b)

FIGURE 2.11 Structure of quartz: (a) showing the helical arrangement of tetrahedra; (b) the view from above showing hexagonal symmetry.

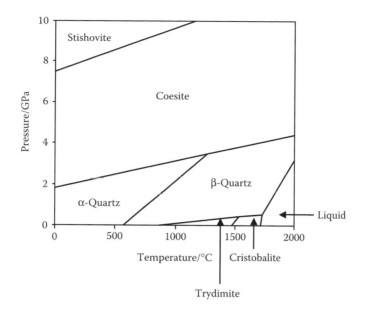

FIGURE 2.12 The phase diagram of silica.

form of SiO_2 called stishovite occurs only rarely because it requires such high pressures to form. One location in which it is found is in the Barringer Meteor Crater in Arizona, where the force of the meteorite impact provided the required pressure. It is built up of linked $[SiO_6]^{8-}$ octahedra rather than tetrahedra because the atoms pack closer together in this arrangement.

The replacement of Si^{4+} by Al^{3+} in framework silicates gives rise to two families of framework aluminosilicate minerals: the **feldspars** and the **zeolites**. In these minerals, the overall charge balance is maintained, as with chain and sheet silicates, by cations such as Na^+, K^+, and Ca^{2+}. The most abundant family of minerals in the Earth's crust is the feldspars, which have the general formula MT_4O_8, where T represents Al or Si in a tetrahedral situation. The end members of the ternary system that account for most of this important family are potassium feldspar (orthoclase, $KAlSi_3O_8$), sodium feldspar (albite, $NaAlSi_3O_8$), and calcium feldspar (anorthite, $CaAl_2Si_2O_8$). Plagioclase feldspar is the name given to minerals having compositions between $CaAl_2Si_2O_8$ and $NaAlSi_3O_8$, and this series of minerals are the commonest rock-forming minerals. These two end members form a solid solution series that involves the coupled substitution of $Na^+ + Si^{4+}$ for $Ca^{2+} + Al^{3+}$. The charge-balancing cations fit into holes or cavities within the framework structure of the feldspar. The ionic radii of Ca^{2+} (114 ppm) and Na^+ (116 ppm) are very similar, so that the three-dimensional structure does not have to distort to accommodate cations of different sizes as Ca^{2+} replaces Na^+ or vice versa. (You may find other values for these ionic radii quoted elsewhere. This is because different texts quote values taken from different sets of radii. The ones used here are those proposed by R.D. Shannon in 1976 and are based on experimentally derived values. Another common set of radii are those of Pauling who used a theoretical model to determine his values.)

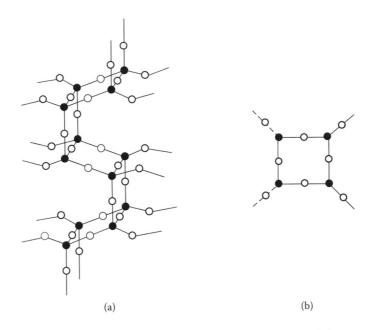

(a)   (b)

FIGURE 2.13 The structure of feldspar: (a) showing the "crankshaft" chain arrangement of tetrahedra; (b) the view from above.

Alkali feldspar is the name given to minerals having compositions between $KAlSi_3O_8$ and $NaAlSi_3O_8$. Unlike the plagoiclase feldspars, solid solution is only possible at high temperatures. In most cases, as an alkali feldspar cools and crystallises, it will separate out into regions that are either high in albite or orthoclase. This can be attributed to the very different ionic radii of Na^+ (116 ppm) and K^+ (152 ppm). This large difference in radius will result in the structure having to distort to accommodate ions of different sizes as Na^+ replaces K^+ or vice versa. Such a distortion will be stable only at high temperatures when the whole lattice has expanded.

Although feldspars are framework aluminosilicates, their structure may be best understood as one built up of linked chains, which in turn are formed from linked squares of tetrahedra. Figure 2.13a shows the chain structure, which is sometimes referred to as a **crankshaft**, and Figure 2.13b shows four tetrahedra linked to form a square that can be stacked to form the chain. Adjacent chains are linked so that every tetrahedron is joined to four others, and this generates the three-dimensional network.

The other general family of framework aluminosilicates are the zeolites, which have much more open framework structures than the compact feldspars. The open structures of the zeolites are reflected in the density ranges of the different framework minerals: zeolites have densities of about 2.1–2.3 g cm^{-3}, feldspars about 2.55–2.76 g cm^{-3}, and quartz has a density of 2.65 g cm^{-3}. The open zeolite structures form channels and cages through which small molecules and ions can move. Most zeolites hold water in these channels and cages, and it is a particular property of zeolites that, when they are heated, they give off water, which can be reabsorbed from the atmosphere. In fact, the name zeolite means "boiling stones."

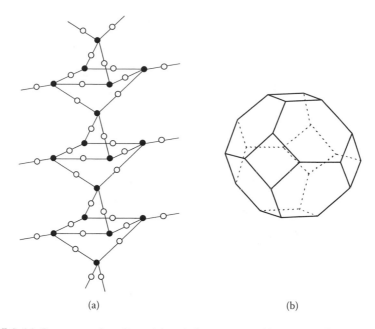

(a)   (b)

FIGURE 2.14 Structures of zeolites: (a) a chain structure with squares of tetrahedra linked by one tetrahedron; (b) the sodalite cage. These can be linked through either the four-ring or the six-ring faces. Each corner represents a T-atom (i.e., Si or Al). Oxygen atoms lie in the middle of the straight lines.

There are many different zeolite structures, and there are whole atlases devoted to describing them; therefore, we mention only a few examples. Their structures are based on a variety of linked chains and cages. In some zeolites, the structure is built up of linked crankshaft chains rather like feldspars, but the linking of the chains is different, giving them a more open structure. An example is phillipsite, $K_2Ca_{1.5}NaAl_6Si_{10}O_{32}.12H_2O$, which occurs widely in seabed deposits. A different chain structure adopted by some zeolites is shown in Figure 2.14a, in which rings of four tetrahedra are linked by a fifth tetrahedron. This is the structure adopted by natrolite, $Na_{16}Al_{16}Si_{24}O_{80}.16H_2O$. Other zeolites are based on cages such as the sodalite or β cage shown in Figure 2.14b. For convenience the atoms are not shown. Each corner represents a T (i.e., either Si or Al) atom, and there is an oxygen atom in the middle of each line. Cages can be joined through the square (four-ring) faces or the hexagonal (six-ring) faces. Zeolites that use this as a structural unit include sodalite itself, $Na_6Al_6Si_6O_{24}.8H_2O$; zeolite A, $Na_{12}Al_{12}Si_{12}O_{48}.27H_2O$; and faujasite, $Na_{12}Ca_{12}Mg_{11}Al_{58}Si_{134}O_{384}.235H_2O$.

One final type of framework silicate we will mention are the magnesium and calcium silicate perovskites, which are thought to be important components of the Earth's lower mantle. In these minerals, which have the general formula $MSiO_3$, the silicon atom is surrounded by six oxygen atoms in an octahedral arrangement. Each octahedron is linked through every corner to six other octahedra, whereas the larger Mg^{2+} or Ca^{2+} ion sits at the centre of eight octahedra, as shown in Figure 2.15. This

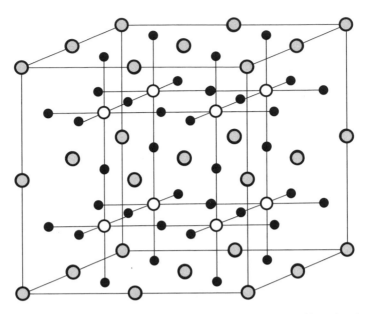

FIGURE 2.15 The structure of magnesium/calcium silicate perovskite, showing silicon atoms (open circles) surrounded by an octahedral arrangement of oxygen atoms (filled dark circles); calcium or magnesium atoms (grey circles) fit between the $[SiO_6]^{8-}$ octahedra.

structure is extremely densely packed, which is why magnesium and calcium silicates adopt it in the lower mantle.

2.3   IGNEOUS ROCKS

Having considered the composition and structure of some common or important types of silicate and aluminosilicate minerals, we can now consider the nature of the **igneous** rocks that are formed from these minerals. However, first, we need to define what is meant by a rock. A rock can be defined as a naturally formed aggregate of one or more types of crystal or grain. A rock that is formed from mineral crystals that have grown together until they are interlocking has a crystalline texture, whereas a rock that is formed of grains or particles that have simply packed together has a fragmental structure. The grains or particles in a rock with fragmental structure may be held together by naturally occurring cement, formed when dissolved minerals such as $CaCO_3$ precipitate out of solution around the grains. Igneous rocks have a crystalline texture, as they are formed from the crystallisation of molten silicate rock.

2.3.1   THE COMPOSITION OF IGNEOUS ROCKS

When we describe the composition of a rock, we can give this in terms of the different minerals forming the rock, the bulk chemical composition of the rock, or the chemical composition of the different minerals. By convention, the abundance of major and minor elements, those which are greater than 0.1% by mass in the rock or mineral, are given as a weight percentage of their oxides, regardless of the

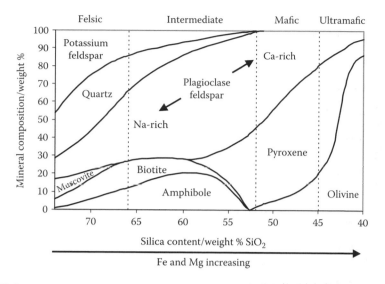

FIGURE 2.16 Classification of silicate rocks by silica composition.

chemical form of the element; for example, the weight percentage of sodium is given as %Na_2O, that of calcium as %CaO, and that of silicon as %SiO_2, even if there is no Na_2O, CaO, or quartz present. The abundance of trace elements, those which are less than 0.1% by mass, are given in parts per million (ppm), where 1 ppm is equivalent to 1 g of element in 10^6 g of rock or mineral (which can also be expressed as 1 mg of the element in 1 kg of rock or mineral).

Igneous rocks can be classified on the basis of %silica (SiO_2) as shown in Figure 2.16, which also shows the approximate mineral compositions of these rocks. **Mafic** rocks have 45–52% SiO_2 by weight and are formed largely of pyroxene and Ca-rich plagioclase (anorthite), with up to about 20% olivine. Because pyroxene and olivine are the main magnesium- and iron-containing minerals, mafic rocks are also rich in magnesium and iron. By contrast, **felsic** rocks are rich in silica, and they have lower concentrations of iron and magnesium because they have much lower proportions of pyroxene and olivine. As a general rule then, %SiO_2 increases with decreasing magnesium and iron. **Intermediate** rocks have 52–66% SiO_2.

SELF-ASSESSMENT QUESTIONS

Q2.1   What is the name given to an igneous rock that is 70% SiO_2?
Q2.2   Give the approximate mineral composition of an intermediate rock with 63% SiO_2.

2.3.2 CRYSTALLISATION OF IGNEOUS ROCKS

Igneous rocks are formed from the crystallisation molten magma. We therefore need to consider the behaviour of rocks as they melt to form a magma and the subsequent crystallisation of the magma. Three pieces of information that we can consider when

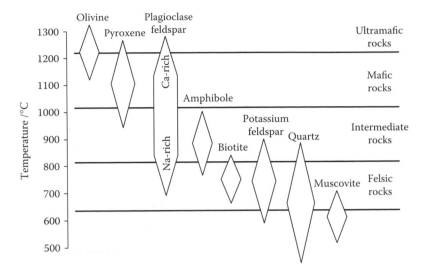

FIGURE 2.17 Crystallisation temperatures of igneous rocks.

working out the crystallisation history of an igneous rock are the texture of the rock, in particular the size of the crystallites; the mineralogy of the rock; and the chemical composition of the rock.

Crystallite size is mainly controlled by the rate of cooling of the molten rock, so that a rock that cools slowly will have a small number of large crystals; i.e., it is **coarse-grained**, whereas a rock that cools very quickly will have a very large number of very small crystallites; i.e., it is **fine-grained**. Sometimes the rate of cooling is very fast and there is no time for crystallites to form, and the rock will have a glassy texture in which the disordered arrangement of atoms in the liquid is retained in the solid. This is the case for the mineral obsidian. Many igneous rocks have large crystals called **phenocrysts** within a very fine-grained **matrix**, or **groundmass**, giving a rock a **porphyritic** texture. This is caused by initial slow cooling of the rock followed by rapid cooling.

Different minerals crystallise at different temperatures, and this is illustrated in Figure 2.17, which shows the approximate temperature ranges over which minerals common in igneous rocks crystallise. Most minerals crystallise over about a 100°C range, but the temperature range for plagioclase feldspars is especially large. As you can see, the crystallisation temperature for plagioclase depends on the relative proportions of $CaAl_2Si_2O_8$ (anorthite) to $NaAlSi_3O_8$ (albite), with the temperature decreasing with increasing proportion of albite. We will be considering the crystallisation of particular minerals and mineral assemblages in more detail in the next section.

SELF-ASSESSMENT QUESTION

Q2.3   How does the crystallisation temperature of different minerals correlate with the approximate level of %SiO_2 in igneous rocks?

When a whole body of magma crystallises, the overall chemical composition of the resulting rock will be the same as that of the original magma. *During* the process of crystallisation, however, the crystals that are forming at any given time will have a different chemical, and therefore mineral, composition compared to that of the parent magma. Therefore, the composition of the remaining magma must be different from that of the original magma composition. As crystals form, they may settle to the bottom or grow on the sides of the magma chamber and will therefore be effectively separated from the crystallising magma. As a result, its composition of the magma will have changed compared to that of the original magma, and therefore, a different assemblage of minerals will crystallise out. This change in rock type as the earlier-formed crystals separate out is called **fractional crystallisation**. The crystals that separate out are called **cumulates**. Crystals that form from the new magma will have a different composition compared to the cumulates because they form from a magma of a different initial composition. Thus, fractional crystallisation results in the formation of igneous rock types different from those one would expect from the parent magma composition.

SELF-ASSESSMENT QUESTIONS

Q2.4   Using Figure 2.16, what rock type will form from the crystallisation of a magma that is 50% SiO_2? If olivine and pyroxene with an average composition of less than 50% SiO_2 crystallises from this magma, what will happen to the silica content of the residual magma?

Q2.5   If the crystals from the magma in SAQ 2.4 separate out when the % SiO_2 of the residual magma has increased to 55%, what rock type will form from the residual magma?

2.3.3 PHASE DIAGRAMS

To study the crystallisation and composition of igneous rocks more closely, we now need to consider the phase diagrams of a number of model systems. The first phase diagram we will consider is that for olivine, which exists as a solid solution series, Mg_2SiO_4 (forsterite, Fo) - Fe_2SiO_4 (fayalite, Fa); see Figure 2.18. This looks rather different from the phase diagrams of carbon dioxide and silica that you have encountered so far. It is an example of a **binary** phase diagram, i.e., a phase diagram for a solid solution in which there are two end members. The composition of the solid solution is plotted on the x-axis, usually as a mass percent or mole fraction of one component, and temperature is plotted on the y-axis. The bottom part of the diagram, below the lower curved line, is the region where the mineral is completely solid for any composition. The lower curved line is called the **solidus** and marks the lowest temperature for each composition where liquid and solid coexist. The middle portion of the diagram, between the two curved lines, is where solid and liquid coexist. The upper curved line is called the **liquidus** and marks the highest temperature for each composition where liquid and solid coexist, and above this the mineral is liquid for all compositions.

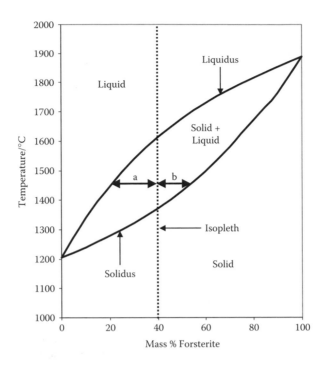

FIGURE 2.18 Phase Diagram of the forsterite–fayalite system.

You can see from the diagram that pure forsterite melts at a much higher temperature than pure fayalite and that intermediate compositions melt at intermediate temperatures. The dotted vertical line is called an **isopleth** and marks a line of overall constant composition for any system. In the region where solid and liquid coexist, the fraction of solid and the fraction of melt at a given temperature are calculated according to Equations 2.1 and 2.2, where *a* and *b* are the distances from the isopleth to the liquidus and solidus, respectively, at that temperature. These can be converted into percentages by multiplying by 100.

$$\text{Fraction of solid} = \frac{a}{a+b} \qquad (2.1)$$

$$\text{Fraction of melt} = \frac{b}{a+b} \qquad (2.2)$$

Consider a melt whose composition is $Fo_{40}Fa_{60}$, at a temperature of 1900°C. This plots in the liquid part of the diagram (Figure 2.19). As the melt cools, olivine crystals start to appear at 1610°C. The composition of the magma at this point is still $Fo_{40}Fa_{60}$. The composition of the first crystals to appear is found by drawing a horizontal **tie-line** from the liquidus at $Fo_{40}Fa_{60}$ to the solidus. The point at which the tie-line crosses the solidus gives the composition of the solid, which in this case

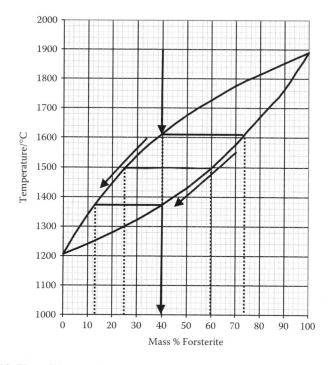

FIGURE 2.19 Phase Diagram for the forsterite–fayalite system.

is $Fo_{74}Fa_{26}$. As you can see, the crystals have a higher proportion of Mg than the melt, and therefore, as the crystals continue to form, the proportion of Fe in the residual melt must increase. As cooling of the melt proceeds, the composition of the melt follows the liquidus, whereas the composition of the crystals follows the solidus. At 1500°C, for example, the remaining melt has a composition of $Fo_{25}Fa_{75}$, whereas the crystals have a bulk composition of $Fo_{60}Fa_{40}$. At 1375°C, virtually all of the melt has crystallised, and has the composition of the original melt, i.e., $Fo_{40}Fa_{60}$, whereas the last remaining trace of melt has the composition $Fo_{13}Fa_{87}$. The whole process can be reversed to determine the composition of the liquid formed on melting crystals of a given composition. You will work through this in SAQ 2.6. The fraction of solid and liquid at 1500°C can be determined as follows. The distance a from the isopleth to the liquidus = 40 – 25 = 15, whereas the distance b from the solidus to the isopleth = 60 – 40 = 20.

$$\text{Fraction of solid} = \frac{a}{a+b} = \frac{15}{35} = 0.43$$

$$\text{Fraction of liquid} = \frac{b}{a+b} = \frac{20}{35} = 0.57$$

In percentages, these are 43% solid and 57% liquid.

SELF-ASSESSMENT QUESTIONS

Q2.6 (i) At what temperature does a solid with the composition $Fo_{70}Fa_{30}$ start to melt?
  (ii) What is the composition of the first liquid produced?
  (iii) At what temperature do the last crystals melt?
  (iv) What is the composition of the last crystals to melt?
  (v) What is the composition of the final melt?
  (vi) What is the percentage of solid and what is the percentage of melt at 1700°C?

If crystallisation occurs sufficiently slowly, there will be continual equilibration between the crystals and residual magma. In the case of olivine, for example, the first crystals formed will be much higher in Mg than the source magma. As the crystal grows, outer layers will have lower levels of Mg. As crystallisation proceeds, however, Mg^{2+} ions will diffuse out from the centre of the crystal to be replaced by Fe^{2+} diffusing into the centre, so that at any temperature the crystals that are present will have the composition expected for that temperature. This is called equilibrium crystallisation.

If crystallisation occurs too quickly for this diffusion process, the chemical composition of the crystals will change as they grow, as there will be insufficient time for equilibrium to be achieved. The outer layers of the crystals will have a different composition from the inner layers (in this case, a higher proportion of Fe and a lower proportion of Mg in the outer layers). The crystals will show **zoning**, with an increasing proportion of Fe on moving out from the centre of the crystal.

Another example of a binary phase diagram involves the plagioclase feldspar solid solution series albite (Ab, $NaAlSi_3O_8$) — anorthite (An, $CaAl_2Si_2O_8$); see Figure 2.20. This is an important solid solution series as plagioclase feldspar is a major constituent in many igneous rocks, and you will investigate it in SAQ 2.7.

SELF-ASSESSMENT QUESTIONS

Q2.7 (i) In the anorthite–albite system, consider a melt at 1550°C that has the composition $An_{50}Ab_{50}$. At what temperature will it start to crystallise?
  (ii) What will be the composition of the first crystals to form?
  (iii) What will be the compositions of the melt and the crystals at 1350°C (assume equilibrium conditions)?
  (iv) What will be the percentages of solid and melt at this temperature?
  (v) At what temperature is crystallisation complete?
  (vi) What will be the compositions of the last drop of melt and the solid at this temperature?

We now move on to a binary phase diagram involving two different minerals, pyroxene (represented by diopside, Di, $CaMgSi_2O_6$) and calcium plagioclase (represented by anorthite), which is shown in Figure 2.21. You can immediately see

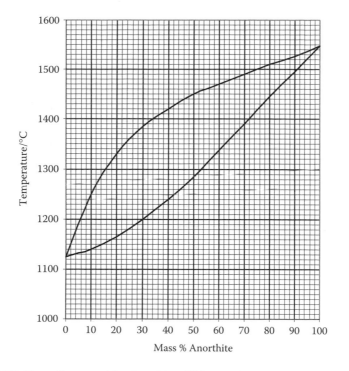

FIGURE 2.20 Phase diagram of the Anorthite–Albite system.

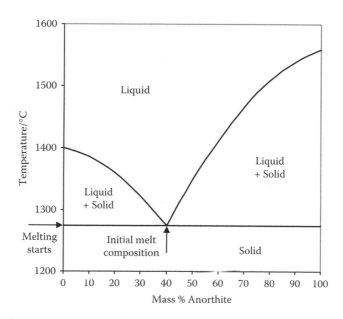

FIGURE 2.21 Phase diagram for the Anorthite–Diopside system.

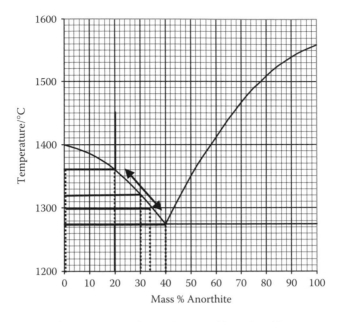

FIGURE 2.22 Crystallisation and melting in the Anorthide–Diopside system.

that the shape of this phase diagram is different from the olivine and plagioclase systems. This is an example of a **eutectic system**. A eutectic is a combination of phases that starts melting at the same temperature regardless of the proportion of each phase and gives a melt of a fixed composition. To investigate the properties of a binary (i.e. involving two phases) eutectic system, we will consider what happens on crystallisation and melting of a sample with the composition $An_{20}Di_{80}$ (Figure 2.22). Consider first a melt of this composition at 1450°C. As it cools, crystallisation will start at 1360°C. The composition of the melt at this point will be $An_{20}Di_{80}$, whereas the composition of the first crystals to form will be An_0Di_{100}, i.e., pure diopside. At 1320°C, the melt will have a composition of $An_{30}Di_{70}$, whereas pure diopside continues to crystallise out. At 1276°C, enough pure diopside has crystallised out for the melt to reach the eutectic composition of $An_{40}Di_{60}$, and this melt now crystallises at this temperature with this composition until all the remaining melt has crystallised.

Let us now see the reverse process. Consider what happens to this material as it is heated from 1200°C. At 1276°C, melting begins, giving a melt whose composition is $An_{40}Di_{60}$. Melting continues at this temperature to give a melt of this composition until all the anorthite has been used up and only pure diopside remains. At this point, the temperature starts to increase as pure diopside starts melting. At 1300°C, sufficient diopside has melted to give the melt an overall composition of $An_{34}Di_{66}$. At 1360°C, there is only a trace of pure solid diopside left, and the melt has an overall composition of $An_{20}Di_{80}$, i.e., the original composition. You will now investigate what happens to a system that has more anorthite than diopside in the eutectic mix.

SELF-ASSESSMENT QUESTIONS

Q2.8   (i)   Consider a melt at 1550°C that has the composition $An_{80}Di_{20}$. At what temperature will crystallisation begin?

      (ii)   What will be the composition of the first crystals to form?

      (iii)   What will be the composition of the melt and the solid at 1400°C?

      (iv)   What happens to the composition of the melt as cooling continues?

      (v)   What happens at 1276°C?

Q2.9   (i)   Consider a solid that has the composition $An_{70}Di_{30}$. At what temperature will melting start?

      (ii)   What is the composition of the melt that is produced?

      (iii)   At what point does the melt start changing composition? What happens to the temperature at this point?

      (iv)   What are the compositions of the melt and the residual solid at 1350°C?

      (v)   At what temperature is melting complete?

      (vi)   What is the composition of the last trace of solid? What is the composition of the melt at this temperature?

2.3.4 TRACE ELEMENTS IN IGNEOUS ROCKS

So far we have considered the bulk composition of various mineral phases. However, igneous rocks also contain trace amounts of many elements, and these can give information about the source of the rock (e.g., did it form directly from mantle material or from melting of previously formed igneous rocks?) and/or its melting or crystallising history (e.g., how much melting of rock took place, and which minerals crystallised first?). Trace elements can be divided between those that are compatible and those that are incompatible. When minerals are crystallising from a melt, compatible elements preferentially go into the crystallising minerals, whereas incompatible elements preferentially stay in the melt. When minerals melt, compatible elements preferentially stay in the mineral, whereas incompatible elements preferentially go into the melt. Compatibility of a trace element M^{n+} in a mineral is measured by the partition coefficient K_d, which is defined by the ratio of the concentrations of M^{n+} in the mineral and in the melt, as shown in Equation 2.3.

$$K_d = \frac{[M^{n+}]_{mineral}}{[M^{n+}]_{melt}} \qquad (2.3)$$

Compatible elements have $K_d > 1$, and incompatible elements have $K_d < 1$. Examples of incompatible elements include K^+, Ba^{2+}, and La^{3+}, whereas examples of compatible elements include Ni^{2+} and Cr^{3+}. It is clear from this that the degree of compatibility does not depend on charge. Rather, it depends on the size of the ion. Ions that are too large will not fit into the structure of the mineral, but ions that are small enough will fit in and replace metal ions in the structure. For example, Ni^{2+}, with an ionic radius of 83 pm, can easily replace Mg^{2+} (86 pm) or Fe^{2+} (92 pm) in olivine and has a K_d of 7 in this mineral, whereas Ba^{2+} (149 pm) is much too large. Similarly, Cr^{3+} (76 pm) can replace Al^{3+} (67.5 pm), but La^{3+} (117 pm) is too large. Ions may also be too small to sit comfortably in a structure. For example, Ni^{2+} has a K_d of 0.01 in plagioclase feldspar. It is much smaller than Ca^{2+} (114 pm) and, therefore, does not replace it because it would "rattle around" in the structure. A few ions are mainly incompatible, but compatible in one or two minerals. For example, Y^{3+} (104 pm) and Yb^{3+} (101 pm) are compatible in garnet where they are just small enough to replace Al^{3+} and still fit into the structure. Remember that, in substituting metal ions, charge balance must be maintained.

SELF-ASSESSMENT QUESTIONS

Q2.10 (i) Sr^{2+} has an ionic radius of 132 pm. It has a K_d in olivine of 0.01. Why?

(ii) Sr^{2+} has a K_d in plagioclase feldspar of 2. Can you account for this K_d value?

2.3.5 MINERAL STABILITY

We have already seen the phase diagram for silica, which shows that the stable polymorph of SiO_2 at atmospheric pressure is quartz. Then, how is it that the other forms can also be observed in nature when, according to the phase diagram, they are not stable at normal temperatures and atmospheric pressures? For example, both of the high-pressure forms, coesite and stishovite, have been observed in the Bar-ringer Meteor Crater, and trydimite and cristobalite are found in volcanic rocks. These other forms are **metastable** forms. Metastable forms are thermodynamically unstable, but the energy required to convert them to the stable form (e.g., the conversion of trydimite to quartz in the case of SiO_2 at normal temperatures and atmospheric pressure) is too high and therefore it does not occur. This is because changing form involves the breaking of chemical bonds. By contrast, the conversion from α-quartz, α-cristobalite, or α-trydimite to β-quartz, β-cristobalite, or β-trydim-ite, respectively, only requires some twisting of the SiO_4 tetrahedra, and this can happen readily.

Box 2.2 Determining the Elemental Composition of Rocks

The chemical composition of a rock depends on the composition of the source rock, the conditions under which the source rock melted, and the subsequent crystallisation history of the rock. It is therefore very useful to be able to determine the exact composition. But how do we determine the chemical com-position of rocks? The composition of a rock is determined by a technique called X-ray fluorescence spectroscopy (XRF). A sample of the rock is bombarded with X-rays. These X-rays excite electrons from low-energy levels to higher-energy levels. As the electrons drop back to their original energy levels, they emit X-rays of characteristic wavelengths. The identity of the different elements present is determined by the wavelengths of X-rays emitted, while the intensity of the emitted X-rays determines how much of each element is present. This technique gives the bulk composition of the rock.

We often wish to determine the elemental composition of each phase within a rock; do this, a technique called scanning electron microscopy (SEM) electron microprobe analysis is used. In SEM, a beam of electrons is used in the same way as a beam of light in light microscopy to produce an image of the rock sample. This image will show the individual grains. Selected locations on the sample (e.g., on different grains) are chosen for electron microprobe analysis to determine the elemental composition. Again, this involves measuring the intensity and wavelength of emitted X-rays to identify and quantify the elemental compositions at those locations. It is even possible to determine how the elemental composition in an individual grain changes, so that zoning can be observed.

2.4 SEDIMENTARY ROCKS

Sedimentary rocks form when rock-forming material is deposited or precipitated, then buried, and consolidated into rock. Most of the material forming sedimentary rocks, about 85%, is derived from the weathering (breaking down) of pre-existing rock. This is then transported by wind or water, often over considerable distances, before being deposited. Other sedimentary rocks are formed from the accumulation of hard parts of marine organisms, or from direct precipitation of minerals from water.

Material that is transported prior to deposition becomes sorted according to the size of the grains. For example, when sediment is being carried by a river, gravels will tend to be deposited upstream, sands will tend to be deposited in the river mouth, and fine silts will tend to be carried out to sea. The greater the distance and the longer the time over which sediment transport occurs, the greater is the extent of sorting of the sediment. A rock that is **well sorted**, i.e., made of very uniformly sized grains, is generally formed from sediment that has been transported over great distances for a long period of time before being deposited, whereas a rock that is **poorly sorted**, i.e., made of grains of very different sizes, is formed from sediment that has been deposited close to its place of formation.

The deposited sediment eventually becomes consolidated into rock. **Diagenesis** is the name given to the various processes that **lithify** the sediment, i.e., convert it into rock; these are described in Table 2.3. As the deposited sediment gets covered by later deposits, it gets compressed, and the rock grains pack together tightly, forcing out water. This may cause some material to dissolve under the pressure of the overlying sediment. Dissolved material precipitates out in spaces or pores in the sediment and also between grains, cementing them together, thus consolidating the loose grains into rock. The cement is very often calcite, $CaCO_3$, or dolomite, $CaMg(CO_3)_2$.

TABLE 2.3
Diagenetic Processes

Diagenetic Process	Description of Process
Compaction	Sediment is compressed by weight of overlying material, forcing out water and reducing pore space between grains
Solution	Increased dissolution of sediment into circulating pore water due to increased pressure
Pressure solution	Increased dissolution due to high pressure at points where grains are in contact with one another.
Recrystallisation	Localised dissolution and recrystallisation of mineral grains to fit pore spaces
Replacement	Progressive conversion of one mineral to another.
Cementation	Crystallisation of dissolved minerals from circulating pore water cementing grains together
Authigenesis	Growth of new minerals within pore spaces, crystallising out from circulating pore water

We will now consider the different minerals and rocks that form sedimentary rocks. These can be divided into two main categories: siliciclastic rocks, which are formed largely from silicate minerals, and non-siliciclastic rocks.

2.4.1 SILICICLASTIC ROCKS

Siliciclastic rocks are rocks formed from clay minerals (layered aluminosilicate minerals with grain sizes of less than 2 μm), quartz, or other silicate minerals such as feldspars or micas (Section 2.2), generally the products of physical, chemical, or biological weathering. We will consider weathering in more detail in Section 2.6, but for now you only need to be aware that physical weathering breaks rocks into smaller pieces, whereas chemical weathering results mainly in the formation of clays, soluble cations and anions, and quartz.

Siliciclastic rocks are composed of grains or **clasts** ranging in size from very fine-grained micrometre- or even smaller-sized particles to metre-sized boulders. Rocks can be classified by grain size according to the size of the most volumetrically abundant grains. The names of the different sedimentary grains and the siliciclastic rock types formed from them are given in Table 2.4. You should note that, in this context, the term *clay* refers to any particles with grain size less than 4 μm, which is different from the definition of clay minerals given at the beginning of this section, but in fact, most clay-sized particles are formed from clay minerals. Claystones are, therefore, formed predominantly from clay minerals, which are the end products of chemical weathering.

Sandstones are formed from sand-sized grains of quartz, feldspar, or small rock fragments. Sandstones, which are composed of more or less 100% quartz grains, are derived from material that has been pretty well completely weathered, whereas sandstones that contain feldspars or rock fragments are derived from incompletely weathered material (see Section 2.6). Conglomerates (in which the rock fragments

TABLE 2.4
Naming of Sedimentary Rock Types by Grain Size

Grain Size	Description	Name of Sediment	Name of Sedimentary Rock	
>256 mm	Very coarse	Boulders	Conglomerate (rounded clasts) or	
64–256 mm		Cobbles	breccia (angular clasts)	
4–64 mm		Pebbles		
2–4 mm	Coarse	Granules		
62.5 μm–2 mm	Fine–medium	Sand	Sandstone	
4 μm–62.5 μm	Very fine–fine	Silt	Siltstone	Mudstone
<4 μm	Very fine	Clay	Claystone	(shale)

are rounded clasts) and breccias (in which the rock fragments are angular clasts) are formed from material that has been only partially weathered.

2.4.2 Carbonates

The most extensive non-siliciclastic sedimentary rocks are carbonates, which form about 20% of the sedimentary cover, averaged over the whole Earth. The principal mineral in carbonate sedimentary rocks is calcite (Chapter 1, Section 1.5.3), which is the main polymorph of calcium carbonate, $CaCO_3$. Aragonite is a less-common polymorph of $CaCO_3$, which is stable at high pressure and forms the hard parts of some organisms.

$CaCO_3$ deposits can form by either chemical deposition when $CaCO_3$ precipitates out of warm, shallow water, or by biological deposition from the accumulated deposits of dead marine animals or algae that have hard parts formed from calcite or aragonite. $CaCO_3$ precipitates out of warm water because the solubility of carbon dioxide, CO_2, decreases with temperature, and this in turn causes soluble HCO_3^- (hydrogen carbonate or bicarbonate) to convert to insoluble $CaCO_3$ and CO_2, as shown in Equation 2.4. This is an example of Le Chatelier's principle because the reaction proceeds to increase the concentration of CO_2 in solution to replace CO_2 that is driven off by the warm water temperatures. This reaction is the reverse of the chemical weathering process of carbonate rocks, which you will learn about in Section 2.6. Evaporation of water from an enclosed basin will also cause $CaCO_3$ to precipitate.

$$2HCO_{3(aq)}^- + Ca_{(aq)}^{2+} \longrightarrow CaCO_{3(s)} + CO_{2(g)} + H_2O_{(l)} \qquad (2.4)$$

The other mechanism whereby carbonates are deposited is by biological deposition. In shallow waters, shells or skeletal material from carbonate-excreting organisms such as corals, mussels, etc., build up on the seabed as these organisms die. $CaCO_3$ is deposited in deep water in the form of microscopic skeletal plates, called **coccoliths,** of tiny marine algae. The thick deposits of chalk limestone covering much of

FIGURE 2.23 Coccolith in chalk limestone.

southern England and northern France, and which form the White Cliffs of Dover, were built up this way. An example of a coccolith in a sample of chalk from Hampshire (England) is shown in Figure 2.23.

In very deep water, $CaCO_3$ redissolves in the reverse reaction of Equation 2.4, as the cold temperatures and rising pressure increase the solubility of CO_2. This reduces the proportion of $CaCO_3$ in the accumulating sediment. The depth at which $CaCO_3$ forms less than 20% of marine sediment is known as the **carbonate compensation depth**.

Some limestones are not pure calcium carbonate but contain significant levels of magnesium. Such rocks are typically formed during diagenesis or after lithification when water containing dissolved magnesium ions, Mg^{2+}, percolates through $CaCO_3$, forming dolomite, $CaMg(CO_3)_2$. A distinguishing feature of dolomite compared to $CaCO_3$ in either of its forms is in its reaction with hydrochloric acid, HCl. Whereas calcite or aragonite react readily with cold, dilute HCl, giving off CO_2 gas, dolomite dissolves only slowly and with little **effervescence** (fizzing). Effervescence is seen when more concentrated HCl is used or the acid is warmed.

Frequently found in association with limestones, and in particular chalk, are irregular-shaped nodules of **chert** (commonly called flint), made of microcrystalline silica, SiO_2. Some marine organisms, e.g., diatoms, which are single-cell algae, and *Radiolaria*, which are moderately large zooplankton, have skeletons based on silica rather than calcium carbonate. As the skeletons accumulate on the seabed and become buried, the silica skeleton is slowly dissolved and reprecipitated in pore spaces to form the chert nodules.

2.4.3 EVAPORITES

Evaporites are mineral deposits that have formed from the evaporation of water in shallow lakes or marine basins. For example, there are evaporite deposits several kilometres thick under the Mediterranean Sea because it has undergone periodic episodes of evaporation followed by inundation with more seawater. Minerals crystallise out from evaporating seawater in the order calcite, gypsum (calcium sulfate,

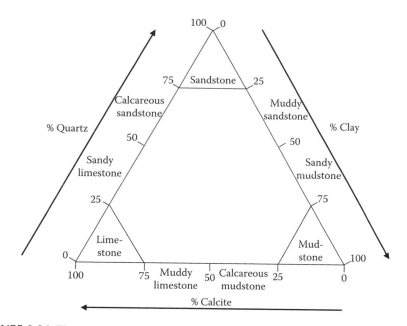

FIGURE 2.24 The sedimentary rock triangle.

$CaSO_4.2H_2O$), halite (sodium chloride, $NaCl$), and then magnesium and potassium salts such as sylvite, KCl. The predominant mineral is halite at about 80% by weight.

The commonest sulfate mineral is gypsum, $CaSO_4.2H_2O$. It forms about 4%, by weight, of the total dissolved solids in seawater, and it occurs both in evaporite deposits and in association with limestones and clays. As you can see from the formula, gypsum has two **waters of crystallisation**, which fit into the structure between the ions. Gypsum can be dehydrated to give anhydrite, $CaSO_4$. Both gypsum and anhydrite occur widely in nature, but gypsum is the initially formed mineral that may later be converted to anhydrite.

2.4.4 THE MINERAL COMPOSITION OF SEDIMENTARY ROCKS

The three components that account for most material in sedimentary rocks are quartz, clays, and calcite. Some sedimentary rocks are composed of largely one mineral, but many sedimentary rocks have significant amounts of two or all three of these components. This can be illustrated in a ternary diagram, as shown in Figure 2.24. The corners of the triangle represent rocks formed of one mineral only; the edges of the triangle represent rocks formed of any two components, and any position inside the triangle represents rocks formed of all three components.

SELF-ASSESSMENT QUESTIONS

Q2.11 What name would you give to a rock that was made of
(i)   70% clay and 30% calcite?

TABLE 2.5
Example Compositions of Sedimentary Rocks

Oxide	Sandstone	Kaolinite	Montmorillonite	Limestone
SiO_2	78.4	44.5	56.6	2.1
TiO_2	0.25	0.15	0.06	0.06
Al_2O_3	4.8	36.6	20.06	0.81
Fe_2O_3	1.1	0.36	3.19	0.54
FeO	0.3	0.07	—	—
MgO	1.2	0.18	3.1	7.9
CaO	5.5	0.19	0.68	42.6
Na_2O	0.45	0.01	2.17	0.05
K_2O	1.3	0.15	0.45	0.33
CO_2	5.0	—	—	41.6
H_2O	1.6	17.4	13.7	0.77
Total	99.9	99.6	100.0	96.8

(ii)   60% calcite and 40% quartz?
(iii) 65% quartz and 35% clay?

2.4.5   The Chemical Composition of Sedimentary Rocks

Sedimentary rocks are usually classified according to their mineral content, average grain size, or both, as already described. Because of the range of compositions of siliciclastic rocks, chemical composition is a much less useful method of classification. However, it does reflect the compositions of the different minerals present. The chemical compositions of a quartz sandstone, two different clay minerals (kaolinite and montmorillonite), and a limestone are given in Table 2.5. By convention, the compositions are given as the oxides of the different elements, even though these may not be present. For example, sodium, Na, is always present as its cation, Na^+, but is given as $\%Na_2O$, and iron is given as iron(III) oxide, Fe_2O_3, although some iron will be present as iron(II). As you can see, quartz sandstone is predominantly composed of silicon in the form of SiO_2, and limestone is predominantly composed of calcium and carbon in the form of $CaCO_3$. Kaolinite and montmorillonite are aluminosilicate minerals, and this is confirmed by the presence of large amounts of both SiO_2 and Al_2O_3 in their chemical compositions.

2.5 METAMORPHIC ROCKS

We have already considered how sedimentary rocks are formed by burial and diagenesis of sediments. Rocks that are buried to greater depths and/or are heated to higher temperatures than required for diagenesis undergo **metamorphism**, which means that there has been a change in the mineral (though not usually chemical) composition of the rocks, or undergo a change in the shape and texture of individual mineral grains and the whole fabric of the rock. During metamorphism, a number of metamorphic processes may take place. The processes that take place depend on the conditions of pressure and temperature to which rocks are subject and the length of time they are subject to them, the particular assemblage of minerals in a given rock being metamorphosed, and the presence or absence of water.

2.5.1 Metamorphism by Recrystallisation

The simplest metamorphic changes take place in rocks that are very largely either pure silica sandstone or pure limestone. In these cases, change takes the form of recrystallisation of SiO_2 or $CaCO_3$, respectively, to form rocks with a crystalline texture, where previously the rock had a fragmental structure. Metamorphosed sandstone forms crystalline quartzite, and metamorphosed limestone forms crystalline marble (now often referred to as metalimestone). The crystalline texture of these recrystallised minerals is often described as "saccharroidal" or "sugary." Dolomite, $CaMg(CO_3)_2$, like calcite, will initially recrystallise to form a marble when heated under pressure, but under more severe conditions it breaks down to form periclase, MgO (Equation 2.5). In the presence of water, brucite, $Mg(OH)_2$, is formed by the hydration of MgO (Equation 2.6), and both species are often observed in metamorphosed dolomites and limestones.

$$CaMg(CO_3)_{2(s)} \longrightarrow MgO_{(s)} + CaCO_{3(s)} + CO_{2(g)} + H_2O_{(l)} \qquad (2.5)$$

$$MgO_{(s)} + H_2O_{(l)} \longrightarrow Mg(OH)_{2(s)} \qquad (2.6)$$

2.5.2 Metamorphism and Chemical Reactions

If limestone contains other minerals, then chemical reactions can take place during metamorphosis, as opposed to the purely physical change involved in recrystallisation. For example, in a limestone containing minor amounts of quartz (SiO_2), calcite reacts with the quartz to form the chain silicate, wollastonite, which has the formula $CaSiO_3$ (Equation 2.7).

$$CaCO_{3(s)} + SiO_{2(s)} \longrightarrow CaSiO_{3(s)} + CO_{2(g)} \qquad (2.7)$$

Rocks formed from clay minerals undergo much more complex changes than sandstone or limestone, involving significant changes in mineralogy as well as rock fabric and grain shape. Differences in the mineral content of the original rock result in the formation of different minerals during metamorphosis. For example, mudstones are fine-grained sedimentary rocks composed largely of clay minerals and quartz. One of the products of metamorphic reaction in mudstones is the aluminsilicate Al_2SiO_5, which does not occur in sedimentary rocks. Al_2SiO_5 together with alkali feldspar is formed in the reaction between muscovite mica and quartz, as shown in Equation 2.8. Figure 2.25 shows a phase diagram for this reaction. To the left of the line (i.e., at lower temperatures), muscovite and quartz are the stable combination; to the right of this line (i.e., at higher temperatures), feldspar and Al_2SiO_5 are the stable combination.

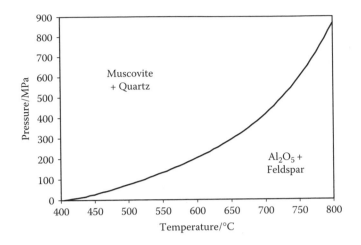

FIGURE 2.25 Phase diagram for the muscovite-quartz system.

$$K_2Al_4[Si_6Al_2O_{20}](OH)_{4(s)} + 2SiO_{2(s)} \longrightarrow 2K[Si_3AlO_8]_{(s)} + 2Al_2SiO_{5(s)} + 2H_2O_{(l)}$$

$$\text{muscovite} \qquad\qquad\qquad\qquad \text{alkali}$$

$$\text{mica} \qquad\qquad\qquad\qquad\qquad \text{feldspar}$$

$$(2.8)$$

Al_2SiO_5 actually exists in three different polymorphs called andalusite, kyanite, and sillimanite. These are stable under different conditions of pressure and temperature, as shown in the Al_2SiO_5 phase diagram in Figure 2.26. The particular polymorph of Al_2SiO_5 formed in this reaction will depend on the pressure and temperature reached during metamorphosis. The phase diagram for Al_2SiO_5 is superimposed on the phase diagram for the muscovite–quartz reaction in Figure 2.27 and shows how Al_2SiO_5 can be used as an **indicator mineral** to put limits on the metamorphic conditions experienced by the rock. The following SAQs explore this system further.

SELF-ASSESSMENT QUESTIONS

Q2.12 Which polymorph of Al_2SiO_5 is stable at 200 MPa and 550°C?

Q2.13 At what temperature will kyanite and sillimanite at a pressure of 600 MPa be in equilibrium?

Q2.14 (i)   At 500 MPa and 500°C, will muscovite and quartz, or feldspar and Al_2SiO_5, be predominant?

     (ii)   At 200 MPa and 250°C, which will be the stable assemblage?

Q2.15 (i)   Given a pressure of 500 MPa, at what temperature will muscovite and quartz be in equilibrium with feldspar and Al_2SiO_5?

     (ii)   Which polymorph of Al_2SiO_5 will be present?

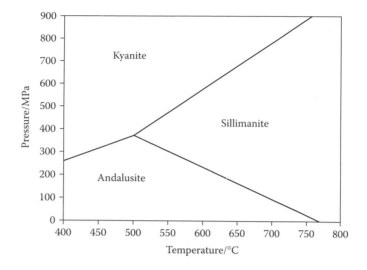

FIGURE 2.26 The andalusite–kyanite–sillimanite phase diagram.

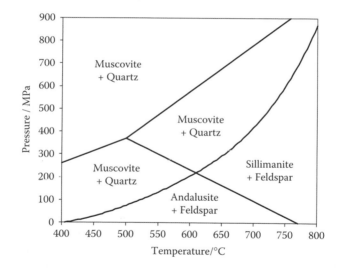

FIGURE 2.27 Al_2O_5 as an indicator mineral in metamorphic rocks.

2.5.3 BIOLOGICAL REACTIONS

A few reactions that cause mineral or chemical change are biological in origin. One example of this is the formation of natural gas and crude oil that you will read about in Box 2.3, and another example is the formation of sulfur deposits. As you have already read, sulfur occurs widely as evaporite deposits of gypsum or anhydrite, but under certain circumstances these can be transformed into deposits of elemental sulfur. Extensive deposits occur, for example, in the United States and Poland. They are usually associated with natural gas or crude oil deposits, which are trapped below evaporite deposits that act as **cap rocks**. Bacteria in the organic layers use sulfate as a source of oxygen, forming hydrogen sulfide, H_2S, as they metabolise hydrocarbons in **anaerobic** reactions (reactions in which oxygen is not used), as shown, for example, in Equation 2.9. Subsequently, hydrogen sulfide undergoes chemical or bacterial oxidation to elemental sulfur (Equation 2.10).

$$CaSO_4 + CH_4 \longrightarrow CaCO_3 + H_2S + H_2O \qquad (2.9)$$

$$8H_2S \xrightarrow{\begin{array}{c}\text{chemical or}\\\text{biological}\\\text{oxidation}\end{array}} S_8 \qquad (2.10)$$

The calcium carbonate that is formed in the initial reduction is found in association with the sulfur. These deposits used to be an important source of sulfur, but they have largely been replaced in recent years by sulfur recovered from natural gas and crude oil, where the sulfur has been removed to cut emissions of sulfur dioxide, SO_2, to the atmosphere.

Box 2.3 The Formation of Fossil Fuels

We have all heard of fossil fuels, but what are they made of and how are they formed? The main fossil fuels are coal, oil and gas, but they also include oil shales, tar sands, and peat. These are all formed from organic matter derived from various types of living organisms that has accumulated in oxygen-poor environments and has been heated under pressure. Fossil fuels are composed mainly of carbon with varying amounts of hydrogen and oxygen. The proportions of the different elements depend on the source material and the conditions to which it was subjected.

Coal is derived from land-based plant material, largely trees, growing in swamps. As the trees died, they accumulated in the swamp water and were rapidly covered by sediments, which excluded oxygen and thus prevented decay. The layers of plant material and sediment subsided as the weight of further plant and sediment layers rapidly built up above them. As the organic matter sank, it heated up. Above about 40°C, volatile organic compounds were driven off, converting the remaining organic matter into lignite, a soft brown coal with about 55% carbon. Above about 70°C, bituminous coal was formed (80% carbon), and above about 170°C, anthracite (90% carbon) was formed. The calorific value of coal — the amount of energy released on burning 1 kg of coal — increases with rising temperatures of the formation, so that anthracite has the highest calorific value. Most of the coal in Britain comes from forests that grew during the Carboniferous period (354–290 million years ago) when Britain was situated near the equator. Peat is plant material that has not got buried quickly enough to get heated up and converted to coal.

Oil and natural gas have the same origin, and almost always occur together in geological formations. Indeed, they are in effect different phases of the same material. They are mainly formed from marine **phytoplankton** (single-celled plants) that sank to the seabed after death and were rapidly buried by sediments. Bacterial decay of the organic material in anoxic conditions resulted in a compact insoluble mass that when heated under pressure, formed a solid waxy substance called kerogen, which is about 75% carbon. The hydrocarbons (compounds composed only of carbon and hydrogen) that form oil (about 85% carbon) and natural gas (75% carbon) were released from kerogen on being heated to between 50 and 200°C (equivalent to burial at depths of between 2 and 7.5 km). Generally speaking, the higher the heating temperature (in effect, the deeper the kerogen was buried), the smaller the carbon chain, so natural gas is the predominant product where heating to over 100°C occurred (equivalent to burial at depths of greater than 3.5 km). Associated with the hydrocarbons are significant levels of sulfur-containing compounds, in particular, hydrogen sulfide, H_2S. This is now recovered from oil and gas and has become the main source of the world's supply of sulfur.

2.6 WEATHERING

As we mentioned in Section 2.3, weathering is a term that covers the various processes that break down and alter rocks. There are two main weathering processes: physical and chemical weathering; however, weathering can also occur through biological mechanisms. We will discuss each type in turn. The extent of physical or chemical weathering depends on the local climate. Chemical weathering is more predominant in warmer climates, whereas physical weathering is more predominant in cooler climates. Some types of weathering also depend on rainfall levels. Rock grains of all sizes that are the products of weathering are transported away and deposited forming sedimentary rocks. These may then be further weathered or metamorphosed in their turn.

2.6.1 PHYSICAL WEATHERING

Physical weathering can be simply described as the breaking up of rock into smaller pieces. Typically, fractures are formed in a rock through a number of mechanisms. Deserts that have high temperature differences between day and night will cause rocks to fracture because of repeated expansion and contraction as the rock heats and cools. A rock will also fracture as a result of the reduction of pressure if it is brought to the Earth's surface, having been previously buried at great depths. The pressure created by mineral crystals that have started growing in cracks will force the sides of the rock to crack apart as the crystal increases in size. When water is present, it can get into small cracks and if the water freezes, it will force the crack further apart because water expands by about 9% on freezing. This freeze–thaw cycle is especially important in cold climates where daytime temperatures reach above 0°C and nighttime temperatures drop well below 0°C, and where there is at least some moderate rainfall to supply the water.

2.6.2 CHEMICAL WEATHERING

Chemical weathering is the breaking up of rock by chemical reaction; we will consider a number of ways in which this can occur. One of the most important chemical weathering reactions is the weathering of carbonate rocks by reaction with rain water containing dissolved CO_2, written as $CO_2(aq)$, as shown in Equation 2.11. This is called carbonation. Rain water is slightly acidic due to the dissolved CO_2 (the theoretical pH of rain water is about 5.6; see Chapter 3, Section 3.3.3), and it is effectively neutralised by reaction with $CaCO_3$. Equation 2.11 is reversible so that solutions containing dissolved hydrogen carbonate will reprecipitate carbonate if the solubility, and hence concentration of CO_2 present in solution, is reduced through heating, or if the concentration of CO_3^{2-} is increased through evaporation of water.

$$CaCO_{3(s)} + CO_{2(aq)} + H_2O_{(l)} \rightleftharpoons Ca^{2+}_{(aq)} + 2HCO^-_{3(aq)} \tag{2.11}$$

At the typical pH of rain water, dissolved CO_2 can be considered as being present in the form of carbonic acid, H_2CO_3, although in fact most of the dissolved CO_2 will simply be present as molecular CO_2 coordinated to H_2O. In Equation 2.11, CO_2 and H_2O were shown as separate compounds, but in the following equations we will show CO_2 dissolved in water as H_2CO_3.

Rain water, as a weakly acidic solution, can cause the chemical weathering of silicate rocks by acid hydrolysis, i.e., by the reaction between water and the rock (which is hydrolysis) in the presence of H^+. (You will learn more about acids in Chapter 3, Section 3.2.) For example, forsterite, the Mg-end member of the silicate mineral olivine, can be hydrolysed, resulting in the formation of soluble Mg^{2+} and silicic acid, H_4SiO_4, as shown in Equation 2.12.

$$Mg_2SiO_{4(s)} + 4H_2CO_{3(aq)} \longrightarrow 2Mg^{2+} + 4HCO^-_{3(aq)} + H_4SiO_{4(aq)} \tag{2.12}$$

The acid hydrolysis of feldspars results in the formation of clay minerals such as montmorillonite and kaolinite, as well as zeolites. The example shown in Equation 2.13 is the acid hydrolysis of potassium feldspar (orthoclase) to form kaolinite. The chemical weathering of feldspars is a very important class of reactions because, as we shall see in Section 2.7, clay minerals are a fundamental component of soils, and without chemical weathering there is no soil formation. Kaolinite in turn can be hydrolysed to bauxite, which is an important ore of aluminium (Equation 2.14).

$$2KAlSi_3O_{8(s)} + 2H_2CO_{3(aq)} + 9H_2O \longrightarrow$$
orthoclase

$$\tag{2.13}$$

$$Al_2Si_2O_5(OH)_{4(s)} + 2HCO^-_{3(aq)} + 2K^+_{(aq)} + 4H_4SiO_{4(aq)}$$
kaolinite

$$Al_2Si_2O_5(OH)_{4(s)} \longrightarrow Al_2O_3.3H_2O_{(s)} + 2H_4SiO_{4(aq)} \tag{2.14}$$
$$\text{kaolinite} \qquad\qquad\qquad \text{bauxite}$$

As you can see in equations 2.11 to 2.13, water is essential for these reactions to occur, and therefore chemical weathering is strongest in wet conditions and does not occur in arid climates. Moderate chemical weathering requires a minimum annual rainfall of about 500 mm, and strong chemical weathering requires rainfall in excess of about 1300 mm. Also, it is a general rule of thumb that the rate of a chemical reaction roughly doubles for every 10°C increase in temperature, and so chemical weathering is stronger in warmer conditions because the rates of the various weathering reactions are faster. One mineral that is very resistant to the chemical weathering reactions described previously is quartz. This is in contrast to simple silicate minerals such as olivine, which dissolves completely, and feldspars, which form clays. Therefore, the ultimate products of chemical weathering of silicate rocks are quartz and clay. As we shall see in the next section, this process is fundamental to the formation of soils. Hydrogen carbonate is formed as part of the weathering process in the reactions described earlier. We will consider the effect of this dissolved hydrogen carbonate on the chemistry of river and sea water in Chapter 3, Section 3.3.

Another general category of chemical weathering reactions is oxidation. You will learn more about oxidation in Chapter 3, Section 3.5, but for now, you only need to understand that a substance is oxidised when it loses electrons or when it gains oxygen. One of the processes in the weathering of iron-containing rocks is the oxidation of Fe^{2+} to Fe^{3+}. In this process, an Fe^{2+} ion, which may also be written as Fe(II) or iron(II), loses an electron to form Fe^{3+}, which may also be written as Fe(III) or iron(III). This oxidation is often observed as a brownish-red staining on rock faces, which is called iron staining, due to the formation of reddish-coloured oxides of Fe^{3+}. The oxidation of fayalite is shown in Equation 2.15, in which Fe^{2+} is oxidised to Fe^{3+}, forming the mineral goethite, $FeO(OH)$. You will sometimes see equations in which the oxidised iron(III) species is given as iron(III) hydroxide, $Fe(OH)_3$, but there is no real evidence for the existence of this species. It is more properly considered either as hydrated ferric oxide, $Fe_2O_3.nH_2O$ (where n is an indeterminate number of water molecules), or as goethite, $FeO(OH)$.

$$Fe_2SiO_{4(s)} + 1/2O_{2(g)} + 3H_2O_{(l)} \longrightarrow 2FeO(OH)_{(s)} + H_4SiO_{4(aq)} \tag{2.15}$$

Sulfide minerals occur widely, both in sedimentary rocks and in veins in igneous and metamorphic rocks, where they frequently occur in association with important commercial minerals that are mined for their metal content. Sulfide minerals oxidise in the presence of air and water by gaining oxygen atoms, producing sulfate, for example, in the oxidation of chalcopyrite, $CuFeS_2$ (Equation 2.16). The copper and iron(II) sulfate minerals thus produced are then further changed, with the formation of copper minerals such as malachite, $Cu_2(OH)_2CO_3$, and the oxidation of iron(II) to iron (III) as ferric oxide or goethite. You will read about the oxidation of iron pyrites, FeS_2 in Chapter 3, Box 3.4.

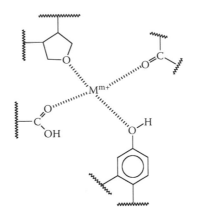

FIGURE 2.28 Chelation of metal ions by organic matter in soil.

$$CuFeS_{2(s)} + 4O_{2(g)} \longrightarrow CuSO_{4(s)} + FeSO_{4(s)} \qquad (2.16)$$

Another mechanism by which rocks can be chemically attacked is by **chelation**. Organic compounds in soil have chemical groups such as carbonyl, $>C=O$, carboxylic acid, $-COOH$ and carboxylate, $-CO_2^-$, which can bind (**chelate**) metal ions, as shown in Figure 2.28, and extract them from the rock surface. Finally, chemical weathering can be as simple as the dissolution of soluble minerals such as halite, NaCl, which are then washed away. The loss of soluble species is termed **leaching**.

2.6.3 BIOLOGICAL WEATHERING

Rocks can be more rapidly weathered through the action of biological systems enhancing physical or chemical weathering. For example, the roots of growing plants can get into tiny cracks and force them apart, which enhances the physical effects of freeze–thaw. Bacteria and lichen can create acidic environments by excreting carboxylic acids. These organic acids increase the rate of chemical weathering by increasing the rate of acid hydrolysis, and they enhance the leaching out of acid-soluble metals. Also, biological activity in soil can increase the concentration and range of organic compounds that are available to chelate metals, extracting them from the rock. The additional acid hydrolysis, leaching, and chelation enhance the chemical weathering of the rock.

2.7 THE CHEMISTRY OF SOIL

Soil is a complex and very variable material. It is formed from three main components: quartz, clay minerals, and organic matter. It is classified on the basis of particle size, with slightly different boundaries compared to those listed in Table 2.4. In a typical soil, the sand-sized (2000–50 μm) and silt-sized (50–2 μm) particles are dominated by quartz (which we have already said is largely inert to chemical weathering), although calcium carbonate may predominate in limestone areas. The clay size fraction (<2 μm) is formed largely from clay minerals (which by definition have a particle size <2μm), humus (which we will discuss in the next section), and metal oxides such as Fe_2O_3.

The three main components of soil occur in differing proportions in different types of soil, and this will depend on the nature of the underlying bedrock, the various weathering processes to which it has been subject, the local vegetation, to what extent material has been washed away or deposited, and what soluble ions have been leached out. However, typically, a soil formed on sandstone bedrock will have a high quartz and low clay content and be very free-draining because of the high proportion of sand-sized particles. The low clay content will mean that the soil will not be able to hold water (remember that clays hold water in their interlayer spaces). Soils formed on chalk or limestone will have high levels of $CaCO_3$.

2.7.1 Soil Organic Matter

The organic matter in soil (soil organic matter, SOM) consists of material from decomposed and decomposing organisms, including plants, fungi and bacteria. For simplicity, we will consider only the contribution of plant material to SOM. Most of the organic matter from plant material is not formed from the kind of simple compounds that we considered in Chapter 1, Section 1.5.1, but is formed from **biopolymers**. A **polymer** is a large compound formed from a large number of simpler molecules (**monomers**) joined together, and a biopolymer is a polymer that occurs in living organisms. The plant component of SOM is derived largely from three biopolymers: cellulose, hemicellulose, and lignin. Cellulose is the main material of plant cell walls, and it consists of many glucose molecules joined together to form long chains (Figure 2.29). Glucose (the monomer) is a simple sugar molecule, and its chemical formula is $C_6H_{12}O_6$; the overall chemical formula of cellulose is usually written as $(CH_2O)_n$ or simply as CH_2O. Hemicellulose has a structure similar to cellulose, having long chains built up from the joining of a variety of simple sugars similar to glucose and related molecules. The third biopolymer of SOM is lignin, which has a much more complex and varied structure than cellulose or hemicellulose and is responsible for the "woodiness" of wood. Its chain structure includes large numbers of aromatic rings, which make it much more resistant to decay, and hence, wooden archaeological artefacts can survive for hundreds of years in **anoxic** (without oxygen) conditions such as waterlogged soils and sediments. An example of a length of lignin chain is shown in Figure 2.30.

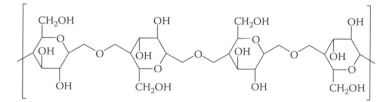

FIGURE 2.29 The structure of cellulose.

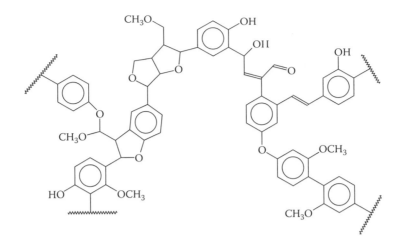

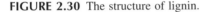

FIGURE 2.30 The structure of lignin.

The ultimate products of decomposition are predominantly CO_2 and water, and the final stage of the decomposition process is often referred to as **mineralisation**. As part of the decomposition process, cellulose is broken down into smaller fragments of 2–3 glucose units, which are then broken down into the glucose monomers and finally oxidised to CO_2 and water. Hemicellulose behaves similarly, and these materials are extensively mineralised within a year. Lignin is much more resistant to decomposition. During the decomposition/mineralisation process, some material undergoes **humification**, which is a process in which organic material becomes more resistant to decay, particularly by the formation of aromatic rings with large numbers of oxygen-containing groups attached. This partially decayed and humified material is called **humus**, which in turn is divided into three fractions: **fulvic acids** (FA), which are water soluble; **humic acids** (HA), which are insoluble in acidic solution; and **humin** (HU), which is insoluble. Fulvic acids are more oxidised than humic acids, with a high proportion of carboxylic acid (-COOH) groups, whereas humic acids have significant proportions of hydroxyl (-OH) groups and carbonyl groups (>C=O). An example of a typical structure of fulvic acid is shown in Figure 2.31.

2.7.2 ION EXCHANGE AND SOIL pH

In this section we are going to consider the pH of soils. You will learn about pH in much more detail in Chapter 3, Section 3.3, but for now you need to know that pH is a measure of the concentration of H^+ ions in solution, and that it is measured on a scale of roughly 0 to 14, where neutral soils have a pH of 7, acid soils have pH values of less than 7 and high concentrations of H^+ ions, and alkaline soils have pH values of greater than 7 and low concentrations of H^+ ions. One very important property of a soil that increases its fertility is its **cation exchange capacity** (CEC). This property is shown by both clay minerals and soil organic matter and is the ability of a material to exchange cations within its structure for cations in solution.

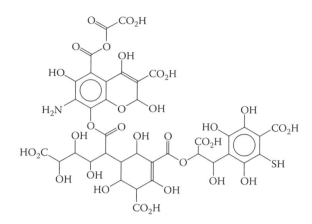

FIGURE 2.31 Structure of fulvic acid.

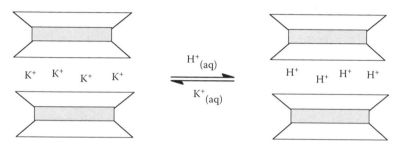

FIGURE 2.32 Cation exchange in clays.

Many clay minerals are formed from negatively charged aluminosilicate layers, which are balanced by cations, typically Na^+, K^+, Ca^{2+}, and Mg^{2+} between the layers (Section 2.2). These cations can be exchanged for ions in water in the soil. In acidic conditions (i.e., a low soil pH) where there is a high H^+ concentration in the soil water, the metal cations are able to exchange for H^+ ions, so that H^+ ions go into the clay layers, and the metal cations are released into solution. This raises the soil pH (decreases the acidity) by removing H^+ ions, thus countering the effect of the acidic conditions. Figure 2.32 shows H^+ ions in solution replacing exchangeable K^+ ions in a clay.

A consequence of the presence of -COOH, -OH, and -C=O groups in SOM is that it adds to the soil's CEC. In neutral and alkaline conditions -COOH groups in soil organic matter dissociate to form negatively charged carboxylate groups -COO⁻, and in alkaline conditions -OH groups dissociate to give phenoxide groups -O⁻. The resulting negatively charged groups are available to take up metal ions from solution (Figure 2.33), and thus, organic matter makes an important contribution to the CEC. H^+ ions are released to soil water in exchange. In soils with high organic matter (e.g., peaty soils that may have up to 90% organic matter), the pH is often very low as there is not enough clay to remove H^+ ions by exchange or carbonate rock to neutralise H^+, and so the soil water has high concentrations of H^+ ions.

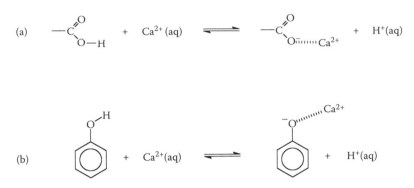

FIGURE 2.33 Cation exchange in organic matter. (a) Cation exchange (e.g. of Ca^{2+}) with organic acids. (b) Cation exchange (e.g. of Ca^{2+}) with phenoxides.

2.7.3 SOIL POLLUTION

Pollution is usually defined as the **anthropogenic** (i.e., by human activity) addition of a substance to the environment above its natural background level that can affect human health or damage ecosystems. Soils can become polluted through a number of routes such as settling or washing of dust and particle grains from the atmosphere, deliberate application of materials onto the soil, e.g., as fertilisers, burial of waste, and as a by-product of industrial or other activity. The aerial route for soil pollution may come through dusts and fumes emitted by vehicles, industrial processes, fires, etc. (see Chapter 4, Section 4.6). However, a particular source that affected much of upland Europe was radioactive fallout from the explosion at the Chernobyl nuclear reactor in 1986.

There are many different soil pollutants, and we cannot cover them all, so we will break the typical soil pollutants down into two basic categories: organic and inorganic. Inorganic pollutants are mostly **heavy metals**, such as cadmium, Cd; mercury, Hg; lead, Pb; zinc, Zn; and copper, Cu. These usually come from steel works, foundries, and metal plating works; various industrial manufacturing processes; waste from mining operations; and road traffic (including run-off), but they can also come from sewage sludge that has been applied to soils. Recent legislation in the European Union (the Directive on the Restriction of Hazardous Substances in Waste Electrical and Electronic Equipment) requires that electrical and electronic equipment manufactured after 1 July, 2006 (with a few exceptions), will no longer be able to contain lead, cadmium, mercury, or hexavalent chromium (i.e., Cr^{VI}), so that the amount of waste containing heavy metals going into landfill sites will be reduced. Heavy metals are positively charged and are, therefore, generally attracted to and held by the negatively charged aluminosilicate sheets in clay minerals in the soil, but especially also by coordination to the various charged and polar groups in soil organic matter (Figure 2.28). Sandy soils, with low levels of clay minerals and organic matter can be a problem as the heavy metals may not be retained in the soil and may consequently leach into ground water. These pollutants generally occur in the ppm to ppb level. The maximum levels of heavy metals in contaminated soils is controlled by EU legislation in Europe and EPA legislation in the U.S., and depends on the land use, e.g., housing, allotments, play areas, industrial uses, or hard landscaping (car parks, etc.).

Some metals such as cadmium can get taken up into plants, where they accumulate, and thus get into the food chain. This is most likely to be a problem where the soil concentrations of the heavy metal are not sufficient to affect the plants (i.e., the metals are at **subphytotoxic** levels), and therefore the problem is not noticed, or where the nature of the soil is such that metals are readily taken up into the plant rather than being retained in the soil. Radioactive inorganic nucleides, including ^{131}I and ^{137}Cs from the 1986 explosion of the Chernobyl nuclear reactor near Kiev, Ukraine (formerly in the Soviet Union), were deposited on land as far away as northern England, Scotland, and Norway. In upland areas where poorer quality soils meant that radioactive nucleides were taken up into plant material in high concentrations, livestock (mainly sheep) that fed on the grass in these areas was banned from sale to prevent them passing into the human food chain. ^{131}I is particularly dangerous as it gets taken into the thyroid gland in the human body where the iodine-containing hormone thyroxine is produced, but the isotope is short-lived ($t_{1/2} = 8.05$ days) and therefore decays rapidly. However, ^{137}Cs has a longer half-life ($t_{1/2} = 30.0$ years), and at the time of writing (2007) is still causing concern, with meat from sheep on some farms continuing to be monitored before being released for sale.

Organic pollutants include polyaromatic hydrocarbons (PAHs), polychlorinated biphenyls (PCBs), pesticides, and herbicides. Many of these pollutants have been designated by the United Nations (U.N.) as persistent organic pollutants (POPs; see Chapter 1, Section 1.5.1). POPs are compounds that are resistant to decomposition (i.e., they are persistent in the environment), are concentrated through the food chain (this is called **bioaccumulation**), and are very toxic; some are known to cause cancer and birth defects, and many others are suspected of similar effects. POPs are regulated by U.N. agreements.

PAHs are serious pollutants because many are, or are strongly suspected of being, **carcinogenic** (cancer-causing), **mutagenic** (they damage DNA), or both. PAHs are formed by incomplete combustion of fuel and are therefore often found in soils by former gas works where gas was obtained by heating coal. They degrade only slowly in the environment because of their extensive aromatic structures (as shown in Chapter 1, Figure 1.33). PCBs are entirely human-made, and similar to PAHs, have been designated as POPs. There are in fact 209 PCBs based on the total number of possible arrangements of chlorine atoms, and two were shown as examples in Chapter 1, Figure 1.34. PCBs were used extensively in electrical equipment such as transformers, and are thus found where such equipment was made, used, or disposed of. They are persistent (slow to degrade) because they are polychlorinated (have several chlorine atoms in their molecular structure), which significantly reduces microbiological decomposition.

Many other major soil pollutants are also highly chlorinated molecules. Two examples are the insecticide DDT and the herbicide 2,4,5,-T (Figure 2.34). DDT, which is banned in the EU but is still used in some parts of the world, has also been designated a POP. It is especially persistent in the environment, having a half-life of 11 years; 2,4,5-T was used by the U.S. military in Vietnam as part of **Agent Orange**. Two groups of chlorinated compounds that are produced when PVC (poly-

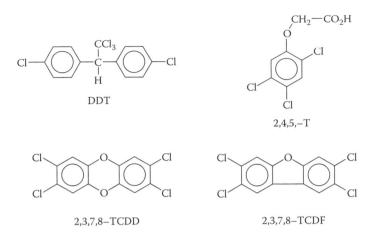

FIGURE 2.34 Some chlorinated organic soil pollutants.

vinylchloride) plastic and similar materials are incinerated at an insufficiently high temperature (<1200°C) are TCDDs (whose structure is based on dibenzo-*p*-dioxin) and TCDFs (whose structure is based on dibenzofuran). The most toxic of these are 2,3,7,8-TCDD and 2,3,7,8-TCDF (Figure 2.34); 3,4,7,8-TCDD is also formed as a by-product in the manufacture of 2,4,5-T, and this led to serious soil contamination in Vietnam. In 1976, an explosion at a chemical works in Seveso, Italy, that was manufacturing a compound called trichlorophenol resulted in severe contamination of the local area by 3,4,7,8-TCDD.

Box 2.4 The Analysis of Polluted Soils

To determine whether a soil is polluted or not, the levels of different pollutants have to be measured, but how is this done? In Box 2.1 we described how the elemental composition of rocks can be determined by XRF analysis, and the same method can be applied to the analysis of heavy metals in soils. This technique, however, is not appropriate for the analysis of organic pollutants in soil, as we want to be able to determine the amounts of specific carbon compounds in the soil, not the total amount of carbon (anyway, carbon is too light an element — it has too low an atomic number — to be determined by XRF analysis). The technique used is called gas chromatography–mass spectrometry (GCMS), but before the sample can be analysed, the organic compounds have to be extracted from the soil sample. This is done by washing the soil sample with an organic solvent that dissolves the organic compounds. The extracted compounds are then injected into the GCMS instrument.

The first part of the instrument — the gas chromatograph — separates the different organic compounds in the extract. The mixture of compounds passes down a 25- to 30-m-long copper coil that is coated on the inside with a high boiling material, usually a type of silicone grease. The coil sits in a heated oven to keep the organic compounds in the vapour phase, and a flow of helium gas carries the compounds through the coil. The compounds adsorb (stick) to the coating, and the stronger they stick to the coating, the longer they take to pass through the coil. This separates the compounds because each different compound adsorbs to a different degree, and therefore takes different lengths of time to pass down the coil. As each compound comes out of the end of the coil it passes into the second part of the instrument — the mass spectrometer. In the mass spectrometer, the molecules of the compound are bombarded with electrons, which causes the molecule to fragment (break up) into ions. The fragment ions are detected and recorded as a *fragmentation pattern*. Each compound has its own characteristic fragmentation pattern, and by comparing the fragmentation pattern of the sample to a computer database, the identity of the different organic compounds in the sample can be determined. The amount of each compound is given by the total number of fragment ions formed, which is compared to the total number of fragment ions in a standard sample of known concentration.

2.8 SUMMARY

In this chapter you should have learned that:

The Earth is 4550 million years old, and it has a core predominantly formed of iron, and a mantle and crust formed largely from silicate minerals.

The main building block of silicate minerals is the silicate group, $[SiO_4]^{4-}$, which is tetrahedral in shape. Silicate groups can link together by sharing oxygen atoms, forming a wide variety of chain, sheet, and three-dimensional network structures.

The negative charge of the silicate group in silicate minerals is balanced by cations such as Ca^{2+}, Mg^{2+}, Fe^{2+}, Na^+, and K^+.

Sheet silicate minerals have, in addition to layers of linked silicate tetrahedra, layers of linked $[AlO_3(OH)_3]^{6-}$ or $[MgO_2(OH)_4]^{6-}$ octahedra. Network minerals such as zeolite and feldspars have $[AlO_4]^{5-}$ tetrahedra in their structures.

Igneous rocks are formed from the crystallisation of molten silicate rock. They can be classified on the basis of their silica composition. We can study the way that they crystallise and melt using phase diagrams. During melting or crystallisation, the melt and solid will have different compositions.

Sedimentary rocks are formed from the physical, chemical or biological deposition of rock material. They are formed largely from quartz, clays and calcite.

Metamorphic rocks are formed when pre-existing rocks undergo a change in their physical form or mineralogical composition caused mainly by heat or pressure. Natural gas and crude oil are formed when the organic remains of marine animals are buried and heated.

Rocks are weathered by physical, chemical, and biological processes. Soil is formed from weathered rock, and organic matter from decomposing organisms, and soil organic matter has a very complex structure.

There are a wide range of soil pollutants, including heavy metals and a variety of persistent organic compounds.

Now try the following questions to test your understanding of the material covered in this chapter:

Q2.16 A particular olivine has 0.6 mol of Mg^{2+} per formula unit. How many moles of Fe^{2+} are there per formula unit? What is the molecular formula of this olivine?

Q2.17 A particular orthopyroxene has 1.3 mol of Fe^{2+} per formula unit. How many moles of Mg^{2+} are there per formula unit? What is the molecular formula of this orthopyroxene?

Q2.18 The formula of tremolite, an amphibole, is $Ca_2Mg_5Si_8O_{22}(OH)_2$. In a related mineral, one Si^{4+} is replaced by an Al^{3+}, with another cation added to balance the charge. Which one of the following three formulas is the correct one?
  (i) $KCa_2Mg_5(Si_7Al)O_{22}(OH)_2$
  (ii) $Ca_3Mg_5(Si_7Al)O_{22}(OH)_2$
  (iii) $Ca_2Mg_5Al(Si_7Al)O_{22}(OH)_2$

Q2.19 Give the approximate mineral composition of a mafic rock that is about 48% SiO_2.

Q2.20 Rubidium, Rb^+ (ionic radius 166 pm), is incompatible in plagioclase feldspar ($K_d = 0.1$). Would you expect Rb^+ to be more compatible or less compatible in potassium feldspar?

Q2.21 Would you expect scandium, Sc^{3+} (ionic radius 89 pm), to be compatible or incompatible in garnet?

3 Water

Water is a unique substance. With a chemical formula of H_2O and a relative molecular mass of only 18.0, the water molecule is deceptively simple, and yet, as far as we can tell, it is essential for life. In fact, the search for life on other planets and planetary satellites has centred on looking for bodies where liquid water may exist now or may have existed in the past. In this chapter you will learn about some of the physical and chemical properties of water that make it such an important molecule, and some of the chemistry relevant to environmental systems.

3.1 THE PROPERTIES OF WATER

So, what are the properties that make water such a unique substance? In this section we will consider first the different phases of water, as solid, liquid, and gas. We will then discuss the polar nature of water and its ability to hydrogen-bond, because this accounts for many of the properties of water, including its behaviour as a solvent. We will also consider how water has profound effects on the Earth's climate at both a global level and on individual weather systems because of its ability to transport heat.

3.1.1 THE PHASE DIAGRAM OF WATER

Water is a liquid at room temperature and atmospheric pressure, but it is present on Earth simultaneously as a solid (in the form of snow, glaciers, polar ice-caps, hail stones, etc.), as a liquid (lakes, rivers, oceans, etc.), and as a gas (atmospheric water vapour). The phase diagram of water is given in Figure 3.1. The dashed horizontal line across the phase diagram marks atmospheric pressure, and you can see that it crosses each of the three fields, confirming that water can exist in all three phases at atmospheric pressure. The point where it crosses the boundary between solid and liquid is the melting point of ice/freezing point of liquid water (at atmospheric pressure), which is 0°C or 273.15 K, and the point where it crosses the boundary between liquid and gas is the boiling point of liquid water (at atmospheric pressure), which is 100°C or 373.15 K. From the slope of the boundary line between the liquid and gas phases, you can see that as the pressure is reduced, which is why the boiling point of water decreases with increasing altitude.

There is only one temperature and pressure at which water exists simultaneously in all three states in the same location. This is known as the *triple point* of water, and is the point on the phase diagram where all three boundary lines intersect, which is at 0.01°C (273.16 K) and 0.006 atm.

One important feature of the phase diagram of water, which is different from that of most other substances, is that the boundary line between liquid and solid has

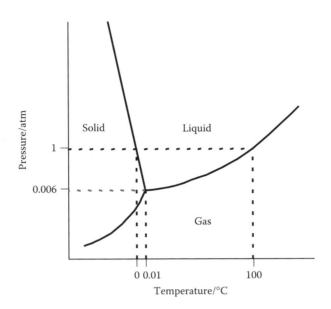

FIGURE 3.1 The phase diagram of water.

a negative slope. What this means is that when pressure is applied to ice that is a few degrees below its melting point, the ice melts. Consider water that is at −5°C and atmospheric pressure. This point lies in the solid stability field and, therefore, water will be in the form of ice. It can be converted to liquid water by crossing the phase boundary between liquid and solid fields either by being heated (moving to the right) or by being subject to pressure (moving upwards). We will discuss the reasons for this unusual behaviour in the next section.

3.1.2 WATER AND HYDROGEN BONDING

As you learned in Chapter 1, Section 1.4.4, water can hydrogen-bond. Each water molecule can form four hydrogen bonds via its two hydrogen atoms and the two lone pairs of electrons on the oxygen, so that every water molecule can be hydrogen-bonded to four other water molecules as shown in Figure 3.2. The number of hydrogens available for hydrogen bonding exactly matches the number of lone pairs, so that every single water molecule can form the maximum of four possible hydrogen bonds. This is the situation in ice and, in fact, the hydrogen bonding acts to hold the water molecules apart in the ice lattice. On melting, about 12% of the hydrogen bonds are broken, allowing the structure to collapse slightly and, hence, liquid water near its freezing point is more dense than ice. The collapse continues as the temperature rises towards 4°C because there is more energy to break hydrogen bonds. As the temperature increases above 4°C, however, the energy gained by the water molecules causes their motion to increase and they move further apart and, therefore, liquid water begins to expand above this temperature, as you would normally expect on heating any liquid or solid. The other aspect of this effect is that when liquid water freezes, it actually expands by about 9%, and therefore ice at 0°C is less dense

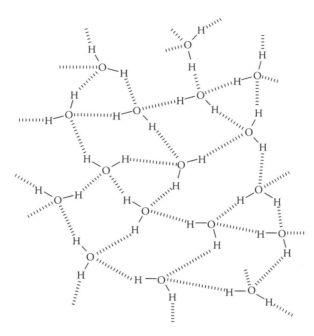

FIGURE 3.2 Hydrogen bonding in water.

than liquid water at 0°C. This is known as the **anomalous expansion of water**, because liquids normally contract on freezing. This anomalous expansion has tremendous implications for life on Earth, and in particular in cold climates, because it means that when water freezes, for example, in a lake, river, or sea, ice forms at the surface, insulating the water below and thus preventing it from freezing. Aquatic animals can survive because the water remains liquid below the ice cover.

The freezing point of seawater is depressed by the salinity (salt content) of the water. This is an example of a **colligative property**, that is, a property that depends only on the number of particles present and not their specific nature. The freezing point of water is depressed because the presence of ions disrupts the formation of the hydrogen bonds that are necessary to hold the water molecules apart. The nature of the different ions does not matter, only the numbers of ions in solution. At the average concentration of dissolved ions in seawater, which is about 35 g kg^{-1}, the freezing point of water is –2°C. As seawater freezes, the ice formed is nearly pure water, while the ions remain in solution. This increases the salinity of the water below the ice that is forming, and this in turn depresses the freezing point of the remaining water still further. Furthermore, as the salinity of the seawater immediately below the forming ice increases, its density increases until it becomes dense enough to sink below the water layers underneath it. This is the source of some of the cold, deep water masses in the Earth's ocean circulation system.

As a result of hydrogen bonding, water has approximately an extra 50 kJ mol^{-1} of energy to hold the water molecules together (four hydrogen bonds, each of about 25 kJ mol^{-1}, with each bond being shared between two molecules), and this explains why water is still liquid at room temperature (25°C) whereas molecules of

TABLE 3.1
Physical Properties of Water and Other Substances

Compound	Formula	Physical State at 25°C	Melting Point/°C	Boiling Point/°C	Specific Heat Capacity/J g^{-1} K^{-1}	Latent Heat of Evaporation/J g^{-1}
Water	H_2O	Liquid	0	100	4.18	2260
Ammonia	NH_3	Gas	−78	−33.6	2.19	1376
Methane	CH_4	Gas	−182	−161.5	2.21	556
Hydrogen fluoride	HF	Gas	−93	20	1.46	374
Hydrogen sulfide	H_2S	Gas	−85	−60	1.0	548
Carbon dioxide	CO_2	Gas	sublimes	−78	0.83	
Carbon disulfide	CS_2	Liquid	−111	46.5	0.99	352
Methanol	CH_3OH	Liquid	−97.5	64.9	2.53	1103
Mercury	Hg	Liquid	−39	357	0.14	295
Sodium	Na	Solid	98	890	1.23	3870

comparable size and RMM such as ammonia, NH_3, methane, CH_4, and hydrogen fluoride, HF, are gases (see Table 3.1). Each molecule of ammonia, NH_3, can form four hydrogen bonds, but on average there can only be two hydrogen bonds per molecule in ammonia, because although there are three hydrogens available on each nitrogen for hydrogen bonding, there is only lone pair of electrons per nitrogen to accept them (Figure 3.3a). Nevertheless, the effect of the two hydrogen bonds per molecule is that ammonia, with an RMM of 17, boils at a much higher temperature than methane (RMM = 16), which is a nonpolar molecule and does not form hydrogen bonds. Hydrogen fluoride (RMM = 20) also forms two hydrogen bonds per molecule, forming a zigzag arrangement of molecules (Figure 3.3b).

The 100°C temperature range over which water is liquid is exceptionally large, and is also due to extensive hydrogen bonding. By comparison, hydrogen sulfide, H_2S (RMM = 34), is liquid over a range of 25°C, NH_3 is liquid over a range of 45°C, and CH_4 is liquid over a range of 21°C (see Table 3.1). Of similar small molecules, only HF has a higher liquid range, and this boils just below room temperature. The difference in temperature range over which these compounds are liquids is also due to the differing abilities of these molecules to hydrogen-bond. Hydrogen sulfide is a slightly polar molecule, and the attraction between the δ− S atoms and δ+ H atoms is an example of a dipole–dipole interaction (Chapter 1, Section 1.4.4), which is weaker than a hydrogen bond.

3.1.3 WATER AND HEAT

Another very important property of water in regard to its role in the environment is its very high **specific heat capacity**. Specific heat capacity, C, is the amount of

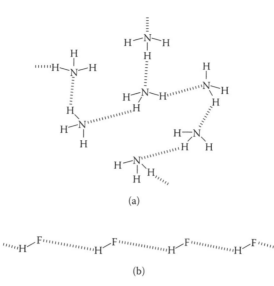

(a)

(b)

FIGURE 3.3 Hydrogen bonding in NH_3 and HF: (a) hydrogen bonding in ammonia, NH_3, showing that there are on average two hydrogen bonds per molecule; (b) showing the zigzag chain of HF molecules owing to hydrogen bonding.

energy that has to be put into a substance to raise its temperature by 1°C. The value for water is 4.18 J g^{-1} °C^{-1}, whereas the value for methanol, CH_3OH, another liquid that is a small molecule, is 2.53 J g^{-1} °C^{-1}, and the value for the element mercury (the only metal that is liquid at 25°C) is 0.14 J g^{-1} °C^{-1}. The low value of C for metals means that the temperature of a metal increases quickly when the metal is heated, and falls quickly when heat is removed. By contrast, the very high value of C for water means that the temperature of water increases slowly when water is heated, but it retains its heat well, and the temperature takes a long time to drop after heating has stopped.

Heat capacity is a very important property for the Earth's environment, because it is the mechanism by which heat is transported around the globe. Water that is warmed in the tropics circulates in the various ocean currents, carrying heat away from the equator to polar regions. The reason why Britain has such a temperate climate for its latitude is that the North Atlantic Current, which is supplied by the Gulf Stream, brings huge quantities of heat from the Caribbean to the Atlantic coast of Western Europe. A quick calculation shows how much energy is transported from the tropics. The flow of surface water carried northwards by the Gulf Stream and North Atlantic Current is about 1.5×10^7 m^3 s^{-1}, i.e., 1.5×10^{13} g s^{-1}. The temperature drop from the tropics to the west coast of Britain is about 17°C, so the amount of energy transported is about 1.1×10^{15} J every second, or about 3.840×10^{18} J per hour (i.e., energy transported per second $= 1.5 \times 10^{13} \times 4.18 \times 17$ J). The total electricity output from all gas-fired power stations in the U.K. was 1.148×10^{18} J over the whole of 2005, so the *hourly* energy transport by the Gulf Stream/North Atlantic Current is over three times greater than the *yearly* electricity output of the U.K.'s gas-fired power stations.

A property related to heat capacity is latent heat of evaporation, which is the energy needed to be put into a substance to convert a liquid at its boiling point to a gas at its boiling point, or the heat given out when a gas at its boiling point condenses to a liquid at its boiling point. The latent heat of evaporation of water is very high, at 2260 J g^{-1}. This is very important because of its effect on weather systems. In particular, it is the huge quantities of energy given out by water vapour as it condenses back to liquid water that drives tropical storms, of which hurricanes (and their Indian Ocean and Pacific Ocean equivalents, cyclones and typhoons) are the most extreme.

3.1.4 WATER AS A SOLVENT

Water is a very good solvent for a wide range of species. These range from simple inorganic ions such as Na$^+$ and Cl$^-$ and small polar organic molecules to much larger polar or charged species and a wide range of physiologically important biomolecules. This versatility as a solvent is because of the polar nature of the water molecule and, especially, its ability to hydrogen-bond. We will consider water as a solvent for inorganic ions in Section 3.3.1; in this section we will consider water as a solvent for organic molecules.

Water is a good solvent for organic molecules that can hydrogen-bond. These are predominantly molecules with -OH, >NH, -NH$_2$, and >C=O groups in their molecular structure. Functional groups that have an attraction for water are called **hydrophilic** ("water-loving"). An organic liquid with hydrophilic groups will be **miscible** with water (i.e., will mix with water), and an organic solid with hydrophilic groups will dissolve in water. An example of a small water-miscible organic liquid is ethanol, CH$_3$CH$_2$OH. On a much larger scale, molecules of dissolved organic matter (including humic and fulvic acids) found in water courses contain large numbers of hydrophilic groups.

Many organic liquids, however, are **immiscible** with water; i.e., they do not mix with water, and many organic solids are insoluble in water. These molecules are **hydrophobic** ("water-hating") because their structures are dominated by nonpolar components such as hydrocarbon chains and rings (chains and rings that contain hydrogen and carbon only). When an immiscible liquid is mixed with water, it will initially form droplets of liquid in the water. This minimises the disruption of the hydrogen bonds between the water molecules. Soon, however, droplets rise to the surface (if the liquid is less dense than water) or sink to the bottom (if the liquid is more dense than water). An example of an immiscible liquid that is less dense than water is crude oil, which is actually a complex mixture of many (mainly hydrocarbon) molecules. In Section 3.6.4 we will consider how oil slicks formed by crude oil spills can be treated.

3.1.5 THE WATER CYCLE

The water cycle, or hydrological cycle, refers to the movement of water between the different **reservoirs** of water on Earth. The total amount of water in the world is about 1380×10^{15} m^3 by volume, or 1.380×10^{21} kg by mass. The approximate amounts (in 10^{15} m^3) and percentages of water in the different reservoirs are given

---

TABLE 3.2
Water Reservoirs

Reservoir	Volume of Water/10^{15} m^3	Percentage of Total/%
Oceans	1344	97.4
Ice caps and glaciers	27.6	2.0
Groundwater and soils	8.3	0.60
Saline lakes	0.097	0.007
Freshwater lakes and rivers	0.097	0.007
Atmosphere	0.014[a]	0.001
Total	1380	100

[a] Volume that atmospheric water vapour would occupy if it were in the liquid phase.

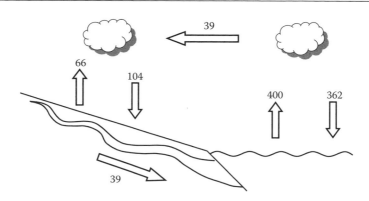

FIGURE 3.4 The water cycle: fluxes through the water cycle in 10^{12} m^3 yr^{-1}.

in Table 3.2. (You may see slightly different values in other books, depending on the source of information.) You can see that by far the biggest reservoir is the oceans, and by contrast, the freshwater essential for our survival constitutes only about 0.007% of the Earth's water. Figure 3.4 shows a very simple diagram with the **fluxes** of water between the different reservoirs, given in 10^{12} m^3 yr^{-1}. The biggest movement is from the oceans to the atmosphere, and most of this water is returned directly to the oceans as rainfall or other precipitation. Similarly, the majority of water that falls on land is lost back to the atmosphere as evaporation. The movement of water vapour in the atmosphere is driven by atmospheric circulation, but also the evaporation and condensation of water is itself also responsible for driving major weather systems because of latent heat, as described in Chapter 1, Section 1.3.1.

3.2 ACIDS, BASES, AND THE pH SCALE

Everyone has heard of **acid** rain as an environmental problem; many people know that an acid is **neutralised** by an alkali or **base**; and many people buy kits to test the **pH** of their soil. But what do the terms *acid*, *base*, *neutralise*, and *pH* mean? An understanding of these concepts is important in many areas of environmental science.

3.2.1 Acids and Bases

There are a number of ways of defining an acid, but one of the most useful was proposed independently in 1923 by J. Brønsted and T. Lowry. The Brønsted–Lowry definition of an acid is **any substance that can donate a proton (i.e., H⁺) to another substance**. Equation 3.1 shows hydrochloric acid, HCl, donating a proton to a hydroxide ion, and Equation 3.2 shows in nitric acid, HNO₃, donating a proton to water.

$$HCl + OH^- \rightleftharpoons Cl^- + H_2O \qquad (3.1)$$

$$HNO_3 + H_2O \rightleftharpoons NO_3^- + H_3O^+ \qquad (3.2)$$

There are a wide variety of compounds that can act as Brønsted–Lowry acids. Those that can donate only one proton, such as HCl and HNO₃, are called **monoprotic** or **monobasic** acids. Some acids, however, are capable of donating two or more protons, and these are called **polyprotic** or **polybasic** acids. An example of a polyprotic acid is phosphoric acid, H₃PO₄. This is a **triprotic** or **tribasic** acid because it can lose three protons. The protons are lost in successive steps shown in Equations 3.3 to 3.5.

$$H_3PO_4 \rightleftharpoons H_2PO_4^- + H^+ \qquad (3.3)$$

$$H_2PO_4^- \rightleftharpoons HPO_4^{2-} + H^+ \qquad (3.4)$$

$$HPO_4^{2-} \rightleftharpoons PO_4^{3-} + H^+ \qquad (3.5)$$

Another example of a polybasic acid is carbonic acid, H₂CO₃, which is a dibasic acid; i.e., it can donate two protons. You will learn more about this important molecule and the whole carbonate system in Section 3.3.3.

In Equations 3.3 to 3.5, the proton, H⁺ is shown existing on its own. In fact, although H⁺ is often shown as a discrete entity in chemical reactions, it always exists in association with another molecule. In aqueous solution, H⁺ is present as the **hydronium ion, H₃O⁺**, in which H⁺ is joined to a molecule of H₂O by sharing a lone pair of electrons from the oxygen atom, as shown in Figure 3.5.

The Brønsted–Lowry definition of a base is **any substance that can accept a proton from another substance**. In Equation 3.1, OH⁻ is the base, and in Equation 3.2, H₂O is the base. The species that is formed by the loss of a proton from an acid (Cl⁻ in

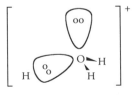

FIGURE 3.5 The hydronium ion.

Equation 3.1 and NO_3^- in Equation 3.2) can itself accept a proton, and is known as the **conjugate base** of the acid. The acid and its conjugate base are known as a **conjugate acid–base pair**, so in Equation 3.1, HCl and Cl⁻ form a conjugate acid–base pair. Similarly, the species formed when a base accepts a proton (H_2O in Equation 3.1 and H_3O^+ in Equation 3.2) is known as the **conjugate acid** of the base. The base and its conjugate acid also form a conjugate acid–base pair, so in Equation 3.1 OH⁻ and H_2O form a conjugate acid–base pair.

There are some compounds that can act as both proton donors and proton acceptors. These species are called **amphiprotic** or **amphibasic**. An example of an amphibasic substance is H_2O. In Equation 3.2 it acts as a base by gaining a proton to give H_3O^+, whereas in Equation 3.6, H_2O acts as an acid by donating a proton to NH_2^-.

$$H_2O + NH_2^- \rightleftharpoons NH_3 + OH^- \qquad (3.6)$$

All the reactions given earlier are shown as being reversible; i.e., they have an arrow showing the reaction going in both the forwards and reverse direction. This is because each species that can gain or lose protons can do this reversibly, so that, for example, H_2O can gain a proton to become H_3O^+, which in turn can lose the proton again.

SELF-ASSESSMENT QUESTIONS

Q3.1 Show how H_2SO_4 can act as a dibasic (diprotic) acid.

Q3.2 Which of the following are amphibasic (amphiprotic) species? Show how they act in this way. (i) H_3PO_4 (ii) $H_2PO_4^-$ (iii) HCO_3^-, and (iiii) CO_3^{2-}.

Q3.3 What are the conjugate bases of (i) HNO_3, (ii) HBr, and (iii) HS⁻?

Q3.4 What are the conjugate acids of (i) CN⁻, (ii) ClO_4^-, and (iii) HS⁻?

3.2.2 The Relative Strength of Acids and Bases

Some acids are better proton donors than others. For example, nitric acid, HNO_3, is a better proton donor than hydrocyanic acid, HCN. In water, HNO_3 will donate virtually all its protons to give H_3O^+ and NO_3^- ions (Equation 3.2), whereas only a small fraction of HCN molecules will donate their protons to form CN⁻ and H_3O^+ ions, and most will remain as HCN (Equation 3.7).

$$HCN + H_2O \rightleftharpoons CN^+ + H_3O^+ \qquad (3.7)$$

Some acids and their conjugate bases are given in Table 3.3. The acids are listed in decreasing order of strength down the table, and the bases are listed in increasing order of strength. Ignore the columns headed K_a and K_b for now. When two acids are in the same solution, the stronger acid will give up its protons before the weaker acid, and the stronger base will be protonated before the weaker base. Consider a solution containing two different acids, e.g., HCl and CH_3CO_2H, and their conjugate

TABLE 3.3
Acid and Base Dissociation Constants

Conjugate Acid	K_a	Conjugate Base	K_b
H_2SO_4	large	HSO_4^-	very small
HCl	large	Cl^-	very small
H_3O^+	55.5	H_2O	1.8×10^{-16}
HNO_3	40	NO_3^-	2.5×10^{-16}
HSO_4^-	1.2×10^{-2}	SO_4^{2-}	8.3×10^{-13}
H_3PO_4	7.9×10^{-3}	$H_2PO_4^-$	1.3×10^{-12}
CH_3CO_2H	1.8×10^{-5}	$CH_3CO_2^-$	5.6×10^{-10}
H_2CO_3	4.2×10^{-7}	HCO_3^-	2.4×10^{-8}
H_2S	1×10^{-7}	HS^-	1×10^{-7}
$H_2PO_4^-$	6.8×10^{-8}	HPO_4^{2-}	1.51×10^{-7}
NH_4^+	5.6×10^{-10}	NH_3	1.8×10^{-5}
HCN	4.0×10^{-10}	CN^-	2.5×10^{-5}
HCO_3^-	4.8×10^{-11}	CO_3^{2-}	2.1×10^{-4}
HPO_4^{2-}	4.4×10^{-13}	PO_4^{3-}	0.023
HS^-	1.3×10^{-13}	S^{2-}	0.077
H_2O	1×10^{-14}	OH^-	1
NH_3	very small	NH_2^-	large

bases, Cl^- and CH_3CO_2H (Equation 3.8). Will the reaction go to the right to give Cl^- and CH_3CO_2H or will the reaction go to the left to give HCl and $CH_3CO_2^-$? HCl lies above CH_3CO_2H in Table 3.3, and therefore HCl is the stronger acid. It will give up its protons more readily than CH_3CO_2H, and therefore Equation 3.8 will go to the right. We get the same answer if we consider the bases: $CH_3CO_2^-$ lies below Cl^- in Table 3.3, and therefore $CH_3CO_2^-$ is the stronger base. It will accept a proton more readily than Cl^-, and therefore Equation 3.8 will go to the right.

$$HCl + CH_3CO_2^- \rightleftharpoons Cl^- + CH_3CO_2H \qquad (3.8)$$

SELF-ASSESSMENT QUESTIONS

Q3.5   What is the conjugate base of H_2SO_4? What is the conjugate acid of NH_3? Write an equation for the reaction between H_2SO_4 and NH_3. Will the reaction go to the right or go to the left?

Q3.6   What is the conjugate base of H_2CO_3? What is the conjugate acid of SO_4^{2-}? Write an equation for the reaction between H_2CO_3 and SO_4^{2-}. Will the reaction go to the right or go to the left?

3.2.3 STRONG ACIDS AND BASES

A **strong acid**, HA, is an acid that is completely or almost completely ionised in water to give H_3O^+ and A^- ions (Equation 3.9). The concentration of H_3O^+ ions produced as the acid ionises is almost equal to the initial concentration of the acid. Sulfuric acid, H_2SO_4, hydrochloric acid, HCl, and nitric acid, HNO_3, are all strong acids because in aqueous solution they are all virtually fully dissociated in water. The dissociation of HNO_3 in water was shown in Equation 3.2. The conjugate base of a strong acid, A^-, is a very **weak** base because it is hardly protonated at all.

$$HA + H_2O \rightleftharpoons A^- + H_3O^+ \qquad (3.9)$$

A **strong base** is a base that is almost completely protonated in water, as shown in Equation 3.10. NH_2^-, which lies below OH^- in Table 3.3, is a strong base because in aqueous solution it will be protonated by H_2O. The reaction between H_2O and amide, NH_2^-, to give hydroxide, OH^-, and ammonia, NH_3, was shown in Equation 3.6.

$$B + H_2O \longrightarrow HB^+ + OH^- \qquad (3.10)$$

3.2.4 Weak Acids and Bases

Most acids are **weak acids**. A weak acid is an acid that is only partially ionised in aqueous solution, as shown in Equation 3.11. The extent of dissociation of a weak acid in water can be quantified using the **acid dissociation constant**, which has the symbol K_a. K_a is calculated from the concentrations of the species present when the acid dissociates as shown in Equation 3.12.

$$HA + H_2O \rightleftharpoons A^- + H_3O^- \qquad (3.11)$$

$$K_a = \frac{[H_3O^+] \times [A^-]}{[HA]} \qquad (3.12)$$

The square brackets are used to indicate concentration; thus, [HA] means "the concentration of HA." [H_2O] is not included in the calculation for K_a, because it is the solvent as well as one of the species in the reaction. Each acid has its own value of K_a that can be looked up in tables of data, and some K_a values are given in Table 3.3. A weak acid has a K_a value of less than 1.

Carbonic acid, H_2CO_3, is an example of a weak acid because in water H_2CO_3 dissociates only to a very small extent to give H_3O^+ and HCO_3^- ions; the reaction lies very much on the left-hand side of the equilibrium (Equation 3.13). A 0.1 M solution of H_2CO_3 will give only 2×10^{-4} M of H_3O^+ ions, i.e., only about 0.2% of H_2CO_3 will be ionised.

$$H_2CO_3 + H_2O \rightleftharpoons HCO_3^- + H_3O^+ \qquad (3.13)$$

A weak base is one that is only partly protonated in water. The strength of a **weak base** can be quantified using the **base dissociation constant**, which is given the symbol K_b. The reaction of a weak base, B, with water is given in Equation 3.14. K_b is determined in the same way as K_a, from the concentrations of the species present, as shown in Equation 3.15. As with K_a values, K_b values can be looked up in tables, and some are given in Table 3.3. A weak base will have a K_b value of less than 1.

$$B + H_2O \rightleftharpoons HB^+ + OH^- \tag{3.14}$$

$$K_b = \frac{[BH^+] \times [OH^-]}{[B]} \tag{3.15}$$

We can use K_a or K_b to calculate the degree of ionisation of a weak acid. Consider, for example, the ionisation of ethanoic acid, CH_3CO_2H (Equation 3.16). If the original concentration of CH_3CO_2H was 0.075 mol dm^{-3}, what will $[CH_3CO_2]$ be, given that the K_a of ethanoic (acetic) acid is 1.8×10^{-5}? We can answer this by making two assumptions. The first is that $[H_3O^+] = [CH_3CO_2^-]$, and therefore $[H_3O^+] \times [CH_3CO_2^-]$ $= [CH_3CO_2^-]^2$. The concentration of un-ionised CH_3CO_2H, $[CH_3CO_2H] = 0.075 -$ $[CH_3CO_2^-]$, but because $[CH_3CO_2^-]$ will be very small, the second assumption we can make is that $[CH_3CO_2H] \approx 0.075$. We can then calculate K_a as shown in Equation 3.17.

$$CH_3CO_2H + H_2O \rightleftharpoons CH_3CO_2^- + H_3O^+ \tag{3.16}$$

$$K_a = 1.8 \times 10^{-5} = \frac{[CH_3CO_2^-] \times [H_3O^+]}{[CH_3CO_2H]} = \frac{[CH_3COO^-]^2}{0.075} \tag{3.17}$$

$$[CH_3CO_2^-]^2 = 1.8 \times 10^{-5} \times 0.075 = 1.35 \times 10^{-6}$$

$$[CH_3CO_2^-] = \sqrt{1.35 \times 10^{-6}} \text{ mol L}^{-1} = 1.16 \times 10^{-3} \text{ mol L}^{-1}$$

Thus, $[CH_3CO_2^-] = 1.16 \times 10^{-3}$ mol dm^{-3}. The percentage ionisation can be calculated very easily as shown in Equation 3.18:

$$\% \text{ ionisation} = \frac{[CH_3CO_2^-]}{[CH_3CO_2H]} \times 100\% = \frac{1.16 \times 10^{-3}}{0.075} \times 100\% = 1.5\% \tag{3.18}$$

For polybasic (polyprotic) acids such as H_3PO_4 that can lose more than one proton, each ionisation step has its own value of K_a. The K_a values get progressively smaller with each ionisation step as it gets harder to remove protons. K_{a1}, the acid dissociation for H_3PO_4 (Equation 3.3), is 7.9×10^{-3}; K_{a2}, the acid dissociation constant for $H_2PO_4^-$ (Equation 3.4), is 6.2×10^{-8}; and K_{a3}, the acid dissociation constant for HPO_4^{2-} (Equation 3.5), is 4.4×10^{-13}.

A more convenient way of expressing K_a values is as pK_a values. These are logarithmic values, which therefore avoids the need for exponents (i.e., where a number is multiplied by a power of 10). The equation linking pK_a and K_a is given in Equation 3.19. pK_a values are typically in the range of about 1 to 16, and because of the minus sign, the larger the pK_a value, the smaller the K_a value.

$$pK_a = -\log_{10} K_a \qquad (3.19)$$

SELF-ASSESSMENT QUESTIONS

Q3.7 (i)   If a 0.2 M solution of HA gives 0.005 M A$^-$ anions, what concentration of H_3O^+ cations will be formed?
      (ii) What concentration of HA will remain un-ionised?
      (iii) What is the percentage ionisation of HA?
      (iv) Calculate the K_a of HA.
Q3.8 Using the value of K_a calculated above, what will [A$^-$] be, approximately, in a 0.05 M solution of HA?
Q3.9 The K_a for H_2CO_3 is 4.2×10^{-7}. What is its pK_a?

3.2.5 THE SELF-IONISATION OF WATER

Water is ionised to a very small extent, producing both H$^+$ (i.e. H_3O^+) and OH$^-$ ions as shown in Equation 3.20. The dissociation constant for this reaction is called the water ionisation constant, K_w, and has the value 1.0×10^{-14} at 25°C (the value changes with temperature). K_w is given by the product of the concentrations of OH$^-$ and H$^+$ (Equation 3.21).

$$H_2O \rightleftharpoons H_3O^+ + OH^- \qquad (3.20)$$

$$K_w = [H^+] \times [OH^-] \qquad (3.21)$$

In absolutely pure water [H$^+$] = [OH$^-$] = 1.0×10^{-7} mol L^{-1}; i.e., there is an equal concentration of H$^+$ and OH$^-$, and it is said to be **neutral**. When an acid is added to water, [H$^+$] increases (i.e., [H_3O^+] increases). Some of these extra H$^+$ cations will combine with OH$^-$ to give H_2O, thus reducing the concentration of OH$^-$ ions, and so in an acidic solution, [H_3O^+]>[OH$^-$]. Conversely, when hydroxide is added to water some OH$^-$ ions will combine with H_3O^+ to produce H_2O, reducing the concentration of H_3O^+ ions, and therefore in a basic solution [OH$^-$]>[H_3O^+]. In each system, the reaction between H_3O^+ and OH$^-$ will proceed until equilibrium is re-established, i.e., until [H_3O^+] × [OH$^-$] = 1.0×10^{-14}. For any conjugate acid–base pair, K_a, K_b, and K_w are connected by Equation 3.22 and, thus, if you know the K_a of a particular acid from tables, you can calculate K_b for its conjugate base.

$$K_w = K_a \times K_b \qquad (3.22)$$

SELF-ASSESSMENT QUESTIONS

Q3.10   In a sample of pure water, if $[H^+] = 4.5 \times 10^{-5}$ mol L^{-1}, what is $[OH^-]$?

Q3.11   (i)   If 0.0040 moles of HCl is added to 250 mL of water, what will the concentration of HCl be in the solution?

(ii)   Assuming that $[H^+]$ already present is negligible compared to the amount of HCl added, what will $[H^+]$ be?

(iii) Assuming that $[H^+]$ already present or used in reaction with OH^- is negligible compared to the amount added, what will $[OH^-]$ be?

Q3.12 K_a for HF is 7.2×10^{-7}. What is K_b for F^-?

3.2.6 THE pH SCALE

When dealing with acids, a more convenient scale for expressing $[H^+]$ (which is in effect the same as $[H_3O^+]$) is the pH scale, which is a logarithmic scale. As with pK_a values, it avoids the need for exponents. The equation linking pH and $[H^+]$ is given in Equation 3.23:

$$pH = -\log_{10}[H^+] \qquad (3.23)$$

Most pH values fall in the range of about 1 to 14. A solution with a pH lower than 7 is **acidic**, whereas a solution with a pH greater than 7 is **basic**, and a solution of pH 7 is neutral. The pH values of some solutions are given in Figure 3.6. Pure rain has a pH of 5.6, and any rain that has a pH less than 5.6 is classed as acid rain. Much acid rain has pH values in the range 4–5, but an example of severe acid rain is one that has occurred in the Sudentic Lakes in Poland, where the pH of the rain has ranged from 2.75 to 3.70. (You will learn more about acid rain in Box 3.1 and Chapter 4, Section 4.6.1.) Even more acidic conditions occur when H_2S gas is converted to sulfuric acid, H_2SO_4, in environments such as some hot springs and caves. Acid mine drainage (see Box 3.4) occurs when metal sulfide minerals are also converted to sulfuric acid, and may have pH values over the range 1–4. The

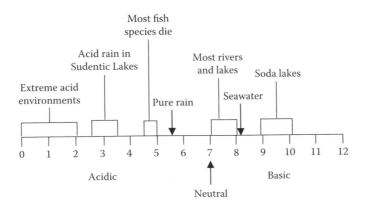

FIGURE 3.6 The pH scale.

pH of seawater is about 8.1, and most lakes and rivers have pH values in the range 7–8, but soda lakes, which have very high concentrations of carbonate ions, CO_3^{2-}, can have pH values up to pH 10. The pH of a solution is usually determined with a pH meter, which gives a digital readout to tenths of a pH unit. pH meters must be calibrated with solutions of known pH before use.

SELF-ASSESSMENT QUESTIONS

Q3.13 In pure water, $[H^+] = [OH^-] = 10^{-7}$ mol dm^{-3}. Show that this gives a pH value for pure water of 7.

Q3.14 The pH of seawater is about 8.1. Calculate $[H^+]$ and $[OH^-]$.

Q3.15 The K_a for H_2CO_3 is 4.2×10^{-7}. What is its pK_a?

3.2.7 ACID–BASE TITRATIONS

The reaction between an acid and a base is often called a **neutralisation reaction** because the pH at the end of a reaction between a strong base and a strong acid is pH 7, as in the reaction between HCl and NaOH (Equation 3.24). One of the products of this reaction is water, which is the case for any neutralisation involving OH$^-$ as the base. The concept of neutralisation in important in an environmental context, for example, in the treatment of land or water affected by acid rain.

$$HCl + NaOH \longrightarrow Cl^- + Na^+ + H_2O \qquad (3.24)$$

As you would expect, the pH of a solution changes as acid is added to base (or vice versa). Figure 3.7 shows how the pH changes as a 0.100 M solution of NaOH is added to 50 mL of a 0.100 M solution of HCl. Initially, the pH changes only slowly as NaOH is added. When nearly 50 mL of NaOH has been added, the pH starts changing very quickly, before levelling out and changing more slowly again. The point at which HCl has been exactly consumed by NaOH is known as the **equivalence point**, and is the midpoint of the marked change. In the reaction of a strong

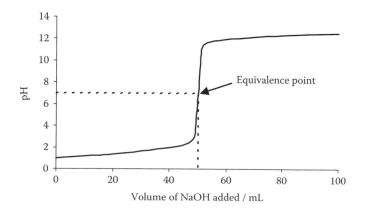

FIGURE 3.7 Titration of HCl with NaOH.

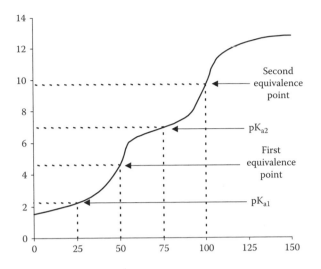

FIGURE 3.8 Titration of H_3PO_4 with NaOH.

acid with a strong base, as in this reaction, the equivalence point occurs at pH 7. When a weak acid reacts with a strong base, for example, the reaction between CH_3COOH and NaOH, the equivalence point occurs at alkaline pHs (i.e., > pH 7). When a strong acid reacts with a weak base, as in the reaction between HCl and NH_3, the equivalence point occurs at acidic pHs (i.e., < pH 7).

The titration curve for a weak, polybasic acid such as H_3PO_4 looks slightly different from that for a strong monobasic acid. The curve for the titration of 50 mL of 0.1 M H_3PO_4 with 0.1 M NaOH is shown in Figure 3.8. Note that there are two equivalence points, which are for the formation of $H_2PO_4^-$, followed by HPO_4^{2-}. The third ionisation stage occurs at a still higher pH. The slope of the curve at the equivalence points are not as steep as in the titration of HCl with NaOH; in other words, the pH does not change as rapidly. The point at which half the acid has reacted with the base is known as the **half-neutralisation point**, and is equal to the pK_a for the acid. H_3PO_4 has three pKa values, of which we can observe the first two, which occur when 25 and 75 mL of NaOH have been added, respectively.

It is possible to determine how much acid is present in a solution by reacting it with a base, or to determine the amount of base in a solution by reacting it with an acid. These types of reactions are called **acid–base titrations**. For the determination of an acid, a known volume of the acid solution to be tested is put into a flask. This is **titrated** with a base whose concentration is accurately known. The base is added in a controlled fashion from a **burette** that measures the volume of liquid added, and the addition continues until the equivalence point is reached. For the determination of a base, a known volume of the base is titrated with an acid of accurately known concentration to the equivalence point. The equivalence point in acid–base titrations is identified by adding a few drops of an **indicator** solution at the start of the titration. An indicator solution changes colour over a known pH range. The large change in pH at the equivalence point causes the indicator to change colour, and at this point the titration is stopped.

At the end of the titration, the volume of acid or base (as appropriate) needed to reach the equivalence point is noted, and from this the concentration of the base or acid solution (as appropriate) can be found. An example is given in the following text for the determination of an NaOH solution by titration with HCl. 43 mL of 0.150 M HCl was needed to neutralise 25 mL of the NaOH solution. The concentration of the NaOH solution is found as follows. First, we calculate the number of moles of HCl used (Equation 3.25):

$$\text{moles HCl} = \frac{\text{concentration HCl} \times \text{volume HCl}}{1000} \tag{3.25}$$

therefore,

$$\text{moles HCl} = \frac{0.150 \times 43}{1000} \text{ mol} = 0.00645 \text{ mol}$$

Then, we need to consider the **stoichiometry** of the reaction between HCl and NaOH. Looking at Equation 3.24, we can see that for every 1 mole of HCl there is 1 mole of NaOH, so the number of moles of NaOH will also be 0.00645 mol. We can now calculate the concentration of the NaOH solution, [NaOH] (Equation 3.26):

$$[\text{NaOH}] = \frac{\text{moles NaOH} \times 1000}{\text{volume NaOH}} \tag{3.26}$$

therefore,

$$[\text{NaOH}] = \frac{0.00645 \times 1000}{25} \text{ mol L}^{-1} = 0.258 \text{ mol L}^{-1}$$

Thus, the concentration of the NaOH solution is 0.258 mol L^{-1}.

SELF-ASSESSMENT QUESTIONS

Q3.16 The concentration of a solution of ethanoic (acetic) acid, CH_3CO_2H, was determined by titration with 0.0500 M NaOH solution. 22.7 mL of NaOH was needed to titrate 50 mL of ethanoic acid. What is the concentration of the ethanoic acid solution?

Q3.17 How many moles of NaOH will be needed to completely neutralise 0.2 moles of H_2SO_4?

3.2.8 BUFFER SOLUTIONS

If a small amount of HCl is added to water, the pH drops markedly from pH 7 to about pH 2, and if a small amount of NaOH solution is added to water, the pH rises markedly from pH 7 to about pH 12. A **buffer solution** by contrast is one in which the pH hardly changes on addition of small amounts of either a strong acid or a strong base. Buffer solutions need to contain two components: a base that will react with added H_3O^+ ions, and an acid that will react with added OH^- ions. The components of the buffer solution must also not react with each other and, therefore, the acid and base are normally an acid–base conjugate pair. Different sets of acid–base conjugate pairs are chosen to give buffer solutions of different pHs. A typical example of a buffer solution is the H_2PO_4/HPO_4^{2-} system. In this system, any added OH^- will react with $H_2PO_4^-$ (Equation 3.27), whereas any added H_3O^+ will react with HPO_4^{2-} (Equation 3.28). A very important natural buffer solution in the environment is seawater. As you will see in the next section, it is a major sink for CO_2, and the buffering effect is due to the presence of the amphiprotic HCO_3^- ion.

$$H_2PO_4^- + OH^- \rightleftharpoons HPO_4^{2-} + H_2O \qquad (3.27)$$

$$HPO_4^{2-} + H_3O^+ \rightleftharpoons H_2PO_4^- + H_2O \qquad (3.28)$$

The pH of a buffer solution is related to the concentrations of the conjugate acid and base making up the buffer and the pK_a of the acid by the **Henderson–Hasselbach equation** (Equation 3.29). Equation 3.30 shows the calculation of the pH of an $H_2PO_4^-/HPO_4^{2-}$ buffer solution where $[H_2PO_4^-]$ (the acid) is 0.500 M, and $[HPO_4^{2-}]$ (the conjugate base) is 0.400 M. K_a for this ionisation step is 6.2×10^{-8}. The pH of this solution is 7.11.

$$pH = pK_a + \log\left(\frac{[\text{conjugate base}]}{[\text{acid}]}\right) \qquad (3.29)$$

$$pH = -\log_{10}(6.2 \times 10^{-8}) + \log\left(\frac{0.400}{0.500}\right) = 7.21 + -0.10 = 7.11 \qquad (3.30)$$

Using the Henderson–Hasselbach equation, it is possible to adjust the pH of a buffer solution to give the desired value by changing the proportions of the acid and its conjugate base (Equation 3.31). In the case of a phosphate buffer, for example, this would mean adding different proportions of KH_2PO_4 and Na_2HPO_4.

$$\frac{[\text{conjugate base}]}{[\text{acid}]} = \text{antilog}\,(pH - pK_a) \qquad (3.31)$$

SELF-ASSESSMENT QUESTIONS

Q3.18   What is the pH of a buffer solution that contains 1.0 M CH_3CO_2H (the acid) and 0.50 M $CH_3CO_2^-$ (the conjugate base)? K_a for this system is 1.8×10^{-5}.

Q3.19 What concentration of HPO_4^{2-} (the conjugate base) will be needed to make a buffer solution of pH 6.85, if the concentration of $H_2PO_4^-$ (the acid) is 0.500 M? K_a for the $H_2PO_4^-/HPO_4^{2-}$ system is 6.2×10^{-8}.

Box 3.1 The Effects of Acid Rain

One consequence of the burning of fossil fuels is **acid rain**, which is rain that has been acidified by dissolved gases such as oxides of sulfur and nitrogen. Normal rain has a pH of 5.6, and acid rain can be defined as any rain with a lower pH than this, with pHs typically in the range pH 4–5, and in some locations as low as pH 2–3. You will learn more about acid rain in Chapter 4, but here we will mention three environmentally important effects of acid rain. The first effect is that of the general problem of decreasing pH on organisms. Most lakes and rivers have pHs in the range 6.5–8.5. If the pH of the lake or river falls below pH 6, the biodiversity is reduced as aquatic species begin to die, and pH is therefore one of the indicators of biodiversity in the aquatic environment. For example, below pH 6 crustaceans die; below pH 5.5 salmon die; and below pH 5.0 pike die.

A second problem caused by acid rain is the mobilisation of otherwise insoluble ions, many of which may be toxic in solution. A particular cause for concern is Al^{3+}, which occurs naturally in clay minerals (see Chapter 2, Section 2.2.4). In acid conditions, Al^{3+} leaches out of the clay and becomes mobile in aqueous solution as $Al^{3+}_{(aq)}$. This is toxic to trees because it gets taken up in the roots and thence into cells, where it binds strongly to a molecule called ATP that controls energy in cells. The liberation of Al^{3+} into soils because of acid rain is thus responsible for the widespread destruction of forests.

A third problem caused by acid rain is damage to buildings and monuments caused by reactions between the stonework and acid. This particularly affects buildings made of limestone as carbonates react with acid to liberate carbon dioxide as shown below:

$$CaCO_{3(s)} + 2H^+_{(aq)} \longrightarrow Ca^+_{(aq)} + CO_{2(g)} + H_2O_{(l)}$$

3.3 IONS IN SOLUTION

In Section 3.1.4 we discussed the role of water as a solvent for organic molecules. In this section we are going to consider water as a solvent of inorganic ions, and the solutions that they form.

3.3.1 THE SOLVATION OF IONS

As you have already learned in Chapter 1, Section 1.3, water is a polar molecule, with the oxygen atom having a partial negative charge, $\delta-$, and the two hydrogen atoms having partial positive charges, $\delta+$. Ions in aqueous solution are attracted to

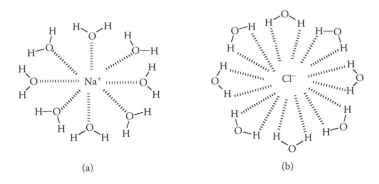

FIGURE 3.9 Solvation of ions showing: (a) a cation (Na$^+$) surrounded by a shell of water, $\delta-$ oxygen atoms pointing in towards the cation; (b) an anion (Cl$^-$) surrounded by a shell of water, $\delta+$ hydrogen atoms pointing in towards the anion.

the oppositely polarised parts of water, so that cations are attracted to the $\delta-$ oxygen atoms, whereas anions are attracted to the $\delta+$ hydrogen atoms. Figure 3.9 shows Na$^+$ and Cl$^-$ ions in water. The Na$^+$ cations are surrounded by $\delta-$ oxygen atoms, and the Cl$^-$ ions are surrounded by $\delta+$ hydrogen atoms. The layers of water molecules that form a shell around an ion are called the **hydration sphere**, and as the ions move through water, they drag their hydration sphere around with them. A small anion such as F$^-$ is able to hold more layers of water around it than a larger anion with the same charge such as Br$^-$ and, similarly, the small cation Li$^+$ is able to hold more water molecules around it than the larger cation K$^+$. Thus, small ions usually have a much larger **hydrodynamic radius** (the radius of the ion plus its hydration sphere) than larger ions of the same charge.

In solution, simple ions such as Na$^+$ and Cl$^-$ are attracted to each other or repelled by each other on the basis of their electrostatic charge, but otherwise they behave as independent entities. However, some oppositely charged ions in solution form specific interactions in which two ions are closely bound together in an **ion pair**. At least one of the ions in an ion pair is usually multiply charged. The ion pair behaves as a single entity, and the ions move around together. An important ion pair in seawater is MgSO$_4$, which is formed from Mg^{2+} and SO$_4^{2-}$ ions.

Many metal ions form **complexes** with small molecules or ions. The molecules or ions around the central metal ion are called **ligands**. In aqueous solution some metals are in the form of **hexaaqua** complexes, in which the metal ion is surrounded by six water molecules arranged as though they are at the corners of an octahedron. Figure 3.10a shows the hexaaqua complex of Fe^{3+}, whose formula is written as [Fe(H$_2$O)$_6$]$^{3+}$. The square brackets in the formula indicate that the metal ion and its associated water molecules form a single unit, and the charge is written outside the square brackets to indicate that the charge belongs to the whole unit. Other examples of hexaaqua ions include [Fe(H$_2$O)$_6$]$^{2+}$ and [Cu(H$_2$O)$_6$]$^{2+}$. Some or all of the water molecules may be replaced by other molecules or ions, as for example in the Fe^{3+} complex [Fe(H$_2$O)$_5$(Cl)]$^{2+}$. In this case, a Cl ion has replaced one water molecule and the overall charge on the complex is +2, i.e., the sum of the charges on Fe^{3+} and Cl$^-$. Actually, the Fe^{3+} hexaaqua complex [Fe(H$_2$O)$_6$]$^{3+}$ only exists at low pH. Above about pH 2, the

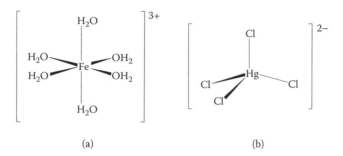

FIGURE 3.10 Metal ion complexes: (a) an octahedral complex of six H_2O molecules around Fe^{3+}; (b) a tetrahedral complex of four chloride anions around Hg^{2-}.

effect of the strong positive charge on Fe^{3+} causes one H_2O molecule to lose H^+, thus forming a hydroxide ion, OH^-. The formula of this complex is $[Fe(H_2O)_5(OH)]^{2+}$.

There may be fewer than or more than six ligands in a complex. In these cases, the arrangements of the ligands around the central metal ion will be different, and an example is shown in Figure 3.10b. In this example, a mercury ion, Hg^{2+}, is surrounded by four Cl^- ions to give a complex with the formula $[HgCl_4]^{2-}$. The chloride ions form a tetrahedral arrangement around Hg^{2+}. This example was chosen because seawater has a high concentration of chloride ions, and this complex is the form that free mercury takes in seawater.

3.3.2 Sparingly Soluble Salts and Solubility Products

In the previous section we considered how ions are held in solution by being attracted to the oppositely charged part of the polar water molecule. But many **salts** (compounds in which the anion is the conjugate base of an acid) are only sparingly soluble, which means that they have solubilities of less than 0.01 mol L^{-1}. Their solubility is governed by their **solubility product**, K_{sp}, which controls the concentrations of the ions that can be in solution before **precipitation** of the sparingly soluble salt occurs. For a salt MA consisting of M^{n+} and A^{n-} ions, K_{sp} is given by Equation 3.32; more generally, for a salt M_nA_m consisting of M^{m+} and A^{n-} ions, K_{sp} is given by Equation 3.33. The units of K_{sp} are mol$^{(n+m)}$ $L^{-(n+m)}$.

$$K_{sp} = [M^{n+}] \times [A^{n-}] \tag{3.32}$$

$$K_{sp} = [M^{m+}]^n \times [A^{n-}]^m \tag{3.33}$$

As an example, consider $Cu(OH)_2$, which consists of Cu^{2+} and OH^- ions. K_{sp} for $Cu(OH)_2$ is 2.0×10^{-19} mol^3 L^{-3} (Equation 3.34). In a solution containing Cu^{2+} and OH^- ions, if the value of $[Cu^{2+}] \times [OH^-]^2$ exceeds 2.0×10^{-19} mol^3 L ^{3}, then $Cu(OH)_2$ will precipitate out of solution, but if $[Cu^{2+}] \times [OH^-]^2$ is lower than 2.0×10^{-19} mol^3 L^{-3}, then $Cu(OH)_2$ will not precipitate out. If $[Cu^{2+}] \times [OH^-]^2$ is lower than 2.0×10^{-19} initially, but then more OH^- ions are added (for example, in the form of NaOH), K_{sp} will be exceeded and $Cu(OH)_2$ will precipitate out of solution.

$$K_{SP} = 2.0 \times 10^{-19} \text{ mol}^3 \text{ L}^{-3} = [Cu^{2+}] \times [OH^-]^2 \qquad (3.34)$$

The effect of the addition of a relatively large amount of one or other of the M^{m+} or A^{n-} ions in the form of another salt leading to a change in the solubility of M_nA_m is called the **common ion effect**. The solubility of M_nA_m can be increased by the addition of large amounts of salts where there is no common ion with M_nA_m. This is known as the **salt effect**. This effect occurs in seawater, where the high concentration of Na^+ and Cl^- ions increases the solubility of $CaCO_3$. You will learn more about the chemistry of seawater in Section 3.4.5.

SELF-ASSESSMENT QUESTION

3.20 (i)   Write out K_{sp} for $BaSO_4$ in terms of the concentrations of Ba^{2+} and SO_4^{2-} ions.

(ii) K_{sp} for $BaSO_4$ is 1.0×10^{-10} mol^2 L^{-2}. If a solution contains 0.05 mol L^{-1} SO_4^{2-} and 3.0×10^{-6} mol L^{-1} Ba^{2+}, will $BaSO_4$ precipitate or stay in solution?

(iii)   If a solution contains 1.0×10^{-6} mol L^{-1} SO_4^{2-} and 3.0×10^{-6} mol L^{-1} Ba^{2+}, will $BaSO_4$ precipitate or stay in solution? If $1\,M$ H_2SO_4 is added to this solution, what will happen?

If the sparingly soluble salt is a salt of a weak acid such as H_2S, then the solubility is pH dependent. Most metal sulfides, for example, cadmium sufide, CdS, mercury sulfide, HgS, and lead sulfide, PbS, are soluble at low pH, and solubility decreases with increasing pH. This can be a problem in the environment if watercourses become acidified through acid rain or other forms of pollution. At the typical pH values of lakes and rivers (~pH 7–8) cadmium sulfide, mercury sulfide, etc., are insoluble and will be held in sediments, but if the water becomes sufficiently acidified, Cd^{2+}, Hg^{2+}, and other toxic ions may be liberated into the water column. We will be considering the effects of pH in more detail in Section 3.5.2 in connection with redox stability diagrams.

3.3.3 THE CARBONATE SYSTEM

The carbonate system is the name given to solutions involving dissolved CO_2 and the carbonate and bicarbonate ions, CO_3^{2-} and HCO_3^-. In effect, this includes most of the Earth's hydrosphere. The carbonate system is, from an environmental point of view, one of the most important chemical systems on Earth. We have already seen how CO_2 gas in the presence of water will weather carbonate rock to form soluble bicarbonate, HCO_3^- (Chapter 2, Section 2.8), and we have seen the reverse reaction in which calcium carbonate precipitates from solutions containing bicarbonate to give carbonate deposits (Chapter 2, Section 2.5). We will now consider the various equilibrium reactions in the carbonate system in more depth.

We will start with the dissolution of CO_2 in water, and the sequence of equilibria that occur as a result. The dissolution of gaseous CO_2 to give $CO_{2(aq)}$ is shown in

Equation 3.35, and the hydrolysis (i.e., the reaction with water) of $CO_{2(aq)}$ to give $H_2CO_{3(aq)}$ is shown in Equation 3.36.

$$CO_{2(g)} \longrightarrow CO_{2(aq)} \tag{3.35}$$

$$CO_{2(aq)} + H_2O_{(l)} \rightleftharpoons H_2CO_{3(aq)} \tag{3.36}$$

As we mentioned in Chapter 2, Section 2.8.2, most dissolved CO_2 is in the form of CO_2 molecules coordinated to water, which is written as $CO_{2(aq)}$, and only a small fraction (about 1/600th of the amount of dissolved $CO_{2(aq)}$) is in the form of H_2CO_3. In fact, the hydrolysis of $CO_{2(aq)}$ is a very slow reaction. However, the equilibria that we are going to consider involve the ionisation of H_2CO_3, and so we will use H_2CO_3 to indicate dissolved CO_2 instead of $CO_{2(aq)}$, just as we did in Chapter 2, Section 2.6.2.

The solubility of CO_2 in water is actually proportional to the partial pressure of CO_2 in the atmosphere, p_{CO_2}, which in turn is directly related to the atmospheric concentration of CO_2 by volume. The relationship between the partial pressure of CO_2 and the solubility of CO_2 (i.e., the concentration of dissolved CO_2), $[H_2CO_3]$, is given by **Henry's Law**, which can be expressed in the form of an equation (Equation 3.37). K_H is known as Henry's constant, and for CO_2 it has a value of 3.4×10^{-2} mol L^{-1} atm^{-1}.

$$K_H = \frac{[H_2CO_3]}{p_{CO_2}} \tag{3.37}$$

The concentration of CO_2 in the atmosphere is about 382 ppmv (parts per million by volume), which is equivalent to a partial pressure of 382×10^{-6} atm. (This was the concentration at the time of writing (2007); the CO_2 level is increasing by about 1–2 ppmv per year.) The solubility of atmospheric CO_2 in water, $[H_2CO_3]$, can then be calculated as shown in Equation 3.38:

$$[H_2CO_3] = K_H \times p_{CO_2} =$$
$$3.4 \times 10^{-2} \times 382 \times 10^{-6} \text{ mol L}^{-1} = 1.3 \times 10^{-5} \text{ mol L}^{-1} \tag{3.38}$$

H_2CO_3 can ionise in two steps (Equations 3.39 and 3.40) to give successively HCO_3^- and CO_3^{2-}. As you can see, at each step one H^+ is released, and therefore water becomes acidic when CO_2 is dissolved in it. The total sum of $CO_{2(aq)}$, $H_2CO_{3(aq)}$, $HCO_3^-{}_{(aq)}$, and $CO_3^{2-}{}_{(aq)}$ is called dissolved inorganic carbon, DIC.

$$H_2CO_{3(aq)} \rightleftharpoons HCO_3^-{}_{(aq)} + H^+_{(aq)} \tag{3.39}$$

$$HCO_3^-{}_{(aq)} \rightleftharpoons CO_3^{2-}{}_{(aq)} + H^+_{(aq)} \tag{3.40}$$

The two ionisation steps can be also given in terms of two acid dissociation constants (Equations 3.41 and 3.42). The acid dissociation constant for the first ionisation step, K_{a1}, is 4.26×10^{-7}, and the acid dissociation constant for the second step, K_{a2}, is 4.68×10^{-11}. These are very small values because H_2CO_3 is a weak acid, and will normally be only very weakly ionised.

$$K_{a1} = \frac{[HCO_3^-] \times [H^+]}{[H_2CO_3]} \tag{3.41}$$

$$K_{a2} = \frac{[CO_3^{2-}] \times [H^+]}{[HCO_3^-]} \tag{3.42}$$

Because the dissolution of CO_2 turns water acidic, rain water is slightly acidic (due to dissolved atmospheric CO_2). We can actually calculate the pH of rain water based on the amount of dissolved CO_2. As you will see shortly, in acidic solutions there is essentially no CO_3^{2-} so that we can ignore the second ionisation step of H_2CO_3 (Equation 3.40), and we only need to consider the first ionisation step, Equation 3.39. This shows us that for every one HCO_3^- formed, one H^+ must be formed, and so we can replace $[HCO_3^-]$ with $[H^+]$ in Equation 3.39 to give Equation 3.43. Equation 3.43 can be rearranged to give an equation in terms of $[H^+]$ (Equation 3.44). We have already calculated $[H_2CO_3]$, and so we can now calculate $[H^+]$. From this we can calculate that the pH of "pure" rain water is 5.6 (Equation 3.45) because of dissolved CO_2.

$$K_{a1} = \frac{[H^+]^2}{[H_2CO_3]} \tag{3.43}$$

$$[H^+] = \sqrt{K_1 \times [H_2CO_3]} = \tag{3.44}$$

$$\sqrt{4.26 \times 10^{-7} \times 1.3 \times 10^{-5}} \text{ mol } L^{-1} = 2.4 \times 10^{-6} \text{ mol } L^{-1}$$

$$pH = -\log[H^+] = -\log(2.4 \times 10^{-6}) = 5.6 \tag{3.45}$$

SELF-ASSESSMENT QUESTIONS

Q3.21 Calculate what the pH of rain water would be if the concentration of CO_2 in the atmosphere were 3750 ppmv, i.e., a partial pressure of 3750×10^{-6} atm, assuming that the overall atmospheric pressure has not changed. (The level of atmospheric CO_2 was much higher earlier in Earth's history than it is today.)

Q3.22 Write out a balanced equation for the reaction between $CO_{2(aq)}$ and $CaCO_3$.

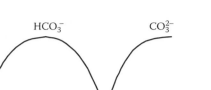

FIGURE 3.11 Carbonate equilibria.

The pH ranges over which H_2CO_3, HCO_3^-, and CO_3^{2-} exist in aqueous solution are shown in Figure 3.11. As we have already mentioned, there is essentially no CO_3^{2-} present in acidic solutions such as rain water, because the presence of additional H^+ ions pushes the equilibrium in both ionisation steps to the left. There is essentially no H_2CO_3 present in basic solution, because excess OH^- ions mop up any H^+ ions formed as H_2CO_3 and HCO_3^- ionise, pushing the equilibrium in both steps to the right. Seawater has an average pH of about 8.1, and therefore CO_2 is mainly in the form of HCO_3^-, with only very small amounts of CO_3^{2-} and very little if any dissolved CO_2 (either as $CO_{2(aq)}$ or H_2CO_3). Most river water has a pH of about 8 owing to the formation of HCO_3^- produced in the reaction between acidic rain water (from dissolved CO_2) and silicate and carbonate rocks (as described in Chapter 2, Section 2.6.2). Seawater and most river water contains HCO_3^- ions, and therefore they are in effect buffer solutions (Section 3.3.8), because HCO_3^- is amphibasic (it can act as a base to mop up acids, and act as an acid to mop up bases). This is why seawater and river water generally have pH values in a very limited range. (Remember that a buffer solution is one in which pH changes only slightly on addition of acids or bases.) The pH values of rivers and streams are much more variable nearer their sources because there are generally much lower levels of HCO_3^- to act as a buffer.

The final component in the carbonate system is the equilibrium between Ca^{2+} and CO_3^{2-} ions in solution and precipitated $CaCO_3$ (Equation 3.46). $CaCO_3$ is, however, a sparingly soluble salt, and therefore it can be considered in terms of the solubility product of $CaCO_3$, where $K_{sp} = 3.8 \times 10^{-9}$ mol^2 L^{-2}. (This is the reciprocal of the value of K used in Section 1.6.3.) Assuming a concentration of Ca^{2+} in seawater of 0.0102 mol L^{-1} (407 mg kg^{-1}), the expected concentration of CO_3^{2-} ions in seawater is 3.7×10^{-7} mol L^{-1}, which is 0.0222 mg kg^{-1}. The actual value is higher, because of the common ion effect; nevertheless, the total concentration of $HCO_3^- + CO_3^{2-}$ ions is 140 mg kg^{-1}, so that you can see carbonate ions form a very small fraction of this total.

$$Ca^{2+}_{(aq)} + CO^{2-}_{3(aq)} \longrightarrow CaCO_{3(s)} \qquad (3.46)$$

A measure of the total amount of carbonate and hydrogen carbonate ions in a solution is the amount of acid needed to convert these ions to H_2CO_3. This is called the **carbonate alkalinity**, A, and is given by Equation 3.47. The concentration of carbonate ions is doubled in the equation because two H^+ are required to neutralise every CO_3^{2-}. The **total alkalinity** is the amount of acid needed to reduce the pH of water to pH 4, which is low enough to protonate the anions of all the weak acids (such as carbonate) in the water. Total alkalinity for river water is essentially equivalent to carbonate alkalinity because $H_2CO_3^-$ is the predominant weak acid. In seawater, some other ions such as $H_2BO_3^-$ make a small contribution to the total alkalinity. You should be careful not to confuse the specific term alkalinity with the more general term **alkaline** (i.e., basic), which means having a pH above 7. The total alkalinity of a sample of water can be determined by titrating it with a strong acid to pH 4. At this pH, all the anions of weak acids are in their protonated forms. The units of alkalinity in Equation 3.47 are mol L^{-1}, i.e., the concentration of acid that is neutralised by HCO_3^- and CO_3^{2-} in the sample. Alkalinity is more often given in units of mg L^{-1} $CaCO_3$, i.e., as the equivalent concentration of $CaCO_3$. To convert alkalinity in mol L^{-1} to alkalinity in mg $CaCO_3$ L^{-1}, multiply by 1000×(molar mass of $CaCO_3$)/2. The factor 2 is because one mole of $CaCO_3$ neutralises 2 moles of H^+.

$$A = \sum [HCO_3^-] + 2 \times [CO_3^{2-}] \qquad (3.47)$$

3.3.4 HARDNESS OF WATER

You are probably familiar with the terms hard water and soft water, and you may have even noticed differences between water in hard-water areas compared to soft-water areas, for example, in the amount of lather formed by soap. **Hardness** is a measure of the amount of dissolved calcium, Ca^{2+}, and magnesium, Mg^{2+}, ions. There are two types of hardness: permanent hardness and temporary hardness. Temporary hardness is due to calcium and magnesium salts that precipitate out of solution on heating, and is primarily due to dissolved calcium carbonate. This type of hardness is particularly prevalent in areas where the geology is predominantly limestone or chalk, and is due to the dissolution of $CaCO_3$ as rainwater seeps through the rock, forming soluble HCO_3^- ions (Chapter 2, Section 2.6.2). When hard water is heated or it evaporates, carbon dioxide is driven off (the solubility of CO_2 decreases with increasing water temperature), precipitating $CaCO_3$ in the reverse of the dissolution reaction (Equation 3.48). Permanent hardness is hardness that is not affected by heating, and is primarily due to the presence of dissolved sulfate and chloride salts of calcium and magnesium such as calcium sulfate, $CaSO_4$, and magnesium chloride, $MgCl_2$.

$$2HCO_3^-{}_{(aq)} + Ca^{2+}_{(aq)} \longrightarrow CaCO_{3(s)} + CO_{2(g)} + H_2O_{(l)} \qquad (3.48)$$

The degree of hardness of water is the total concentration of Mg^{2+} + Ca^{2+} ions present, given in units of in mg L^{-1}, expressed as the concentration equivalent of

CaCO₃ also in mg L⁻¹. The example below shows how to calculate the alkalinity for
a sample of water in which $[Ca^{2+}] = 36.0$ mg L⁻¹ and $[Mg^{2+}] = 12.0$ mg L⁻¹. The
concentration of Mg^{2+} is converted to an equivalent concentration of Ca^{2+} and added
to the actual concentration of Ca^{2+} to give a total concentration of Ca^{2+} (Equation
3.49), and this is then converted to the concentration equivalent of CaCO₃ to give
the hardness (Equation 3.50).

$$\text{Total } [Ca^{2+}] = [Ca^{2+}] + [Mg^{2+}] \times \frac{\text{RAM Ca}}{\text{RAM Mg}} =$$

$$36.0 + 12.0 \times \frac{40.01}{24.31} \text{ mg L}^{-1} = 55.7 \text{ mg L}^{-1}$$

(3.49)

$$\text{Hardness} = [Ca^{2+}] \times \frac{\text{RMM CaCO}_3}{\text{RAM Ca}} =$$

$$55.7 \times \frac{100.02}{40.01} \text{ mg L}^{-1} = 139.2 \text{ mg L}^{-1}$$

(3.50)

3.3.5 THE CHEMISTRY OF SEAWATER

Whereas the composition of river water is very variable, depending heavily on the
geology through which the river runs and the extent of weathering, the composition
of seawater is much more uniform. Table 3.4 gives average values for the concentrations of the 12 major ions in seawater in g kg⁻¹, and you can see that much of
the largest contribution to the major dissolved ions is due to Na^+ and Cl^- (as you
would probably expect), with smaller amounts of Mg^{2+}, Ca^{2+}, K^+, and SO_4^{2-}.

TABLE 3.4
Composition of Seawater

Ion	Concentration/mg kg⁻¹	Concentration/‰
Cl^-	19,330	19.33
Na^+	10,720	10.72
SO_4^{2-}	2,690	2.69
Mg^{2+}	1,286	1.286
Ca^{2+}	407	0.407
K^+	385	0.385
$HCO_3^- + CO_3^{2-}$	140	0.14
Br^-	66	0.066
$H_2BO_3^-$	24	0.024
Sr^{2+}	10	0.01
H_4SiO_4	7	0.007
F^-	1	0.001

The **salinity** of seawater is more or less equal to the average concentration of dissolved ions expressed in ‰ ("per mil"), which is equivalent to g kg^{-1}, and for the values given in Table 3.4 this comes to 35.1 ‰. Salinity, is however, formally defined in terms of the ratio of the conductivity (i.e., the ability to conduct electricity) of the seawater sample to the conductivity of a standard solution of potassium chloride, KCl, at 15°C. The average salinity of seawater is about 34.5 to 35. (Salinity has no units because it is based on a ratio.)

Seawater is an important sink for atmospheric CO_2 because the upper layers of seawater contain a large amount of undissolved $CaCO_3$. This reacts with dissolved CO_2 in the same way that carbonate rocks are weathered (Equation 2.9) to give soluble HCO_3^-. In fact, CO_2 is much more soluble in seawater than in rain water, because of the presence of HCO_3^- and other ions in seawater. One concern about the increasing level of atmospheric CO_2 is that more CO_2 will dissolve in seawater as atmospheric CO_2 levels increase (obeying Henry's law), and the increasing concentration of CO_2 in the oceans will lead to a slow acidification of seawater. This in turn will make $CaCO_3$ more soluble so that corals and other organisms that have skeletons or other hard parts (called tests) made of $CaCO_3$ will suffer as they find it harder to grow their $CaCO_3$ structures. The average pH of seawater is about pH 8.1.

The concentration of dissolved oxygen as O_2 gas varies as oxygen solubility is very temperature dependent. O_2 has a maximum of concentration of about 10.1 mg L^{-1} in freshwater at 15°C, dropping to about 8.6 mg L^{-1} at 25°C, and about 7.0 mg L^{-1} at 35°C. The solubility in seawater is lower, being about 6.7 mg L^{-1} at 25°C. As you can see, the concentration of dissolved oxygen increases as the water temperature decreases (this is also the case for CO_2), so that cold water near the bottom of the oceans is more highly oxygenated than warm surface waters.

Dissolved oxygen levels can also be given as a percentage saturation, i.e., as a percentage of the maximum saturation expected for a given temperature. For example, a freshwater sample with O_2 levels of 8.3 mg L^{-1}, would be 82% saturated if the water temperature was 15°C, but 96% saturated if the water temperature was 25°C.

Box 3.2 CO$_2$ and Killer Lakes

In 1986, 1800 people living near Lake Nyos in Cameroon died in mysterious circumstances. There had been another similar tragedy by Lake Monoun also in Cameroon. After extensive investigation by geologists and others, the cause was discovered: a sudden and dramatic release of carbon dioxide from the lake. Carbon dioxide is a lethal gas. If you were to remain in a sealed room with no ventilation, you would be dead long before the oxygen was used up. What the geologists discovered was that CO_2 was bubbling into the bottom of the lake, coming from geologically active faults in this part of Africa's Rift Valley. The high pressures at the bottom of these deep lakes caused CO_2 to dissolve readily. But just like a bottle of carbonated drink, if the pressure is released, then CO_2 comes rapidly out of solution. This happened when the lake experienced what is called **lake overturn**, in which the water at the bottom of the lake is brought to the surface. In the Lake Nyos tragedy, a landslide triggered the overturn when possibly hundreds of tons of rock and mud slid down the steep slopes of the lake sides into the bottom of the lake. This forced bottom water upwards, and as it did so huge amounts of dissolved CO_2 came bubbling out of solution. CO_2 is denser than air, and the gas literally flowed through the nearby valleys, killing people as they slept.

Geologists are working on an ingenious solution to overcome this problem — using fountains to bring water up from the deep, allowing the water to degas steadily, and preventing the buildup of deep water with very high dissolved CO_2 levels.

Box 3.3 Water Softeners and How they Work

Two particular problems associated with hard water are the deposition of calcium carbonate in water pipes and on heating elements (i.e., the effects of temporary hardness), and the formation of scum when using soap. Soap is formed from a chemical called sodium stearate (stearate is a long-chain organic molecule with a negatively charged group at one end that is balanced by Na^+), and scum is insoluble calcium stearate, which is formed when soap is used in hard water. These two problems of carbonate deposition and scum formation can be overcome with the use of water softeners. But how do water softeners work?

There are two main mechanisms for softening water: **ion exchange** and **chelation**. In a typical ion exchange water softener, an insoluble solid such as a zeolite (see Chapter 2, Section 2.2.5) is used. The zeolite contains Na^+ cations that exchange for Ca^{2+} ions in the water, two Na^+ going into solution for every Ca^{2+} going into the zeolite. Sodium salts are soluble, and so Na_2CO_3 stays in solution. Precipitation of dissolved $CaCO_3$ does not occur, as Ca^{2+} has been removed. This system is used in some domestic water supplies by having the water supply pass through a cartridge of zeolite. The water softener is regenerated periodically by adding common salt, NaCl, to replace the Ca^{2+} ions that have accumulated in the zeolite. Zeolites are also added in significant quantities to washing powders (but not liquid detergents) to prevent the formation of scum by removing Ca^{2+} ions from solution during the wash, exchanging them as before for Na^+ ions.

The second type of water softener acts by forming a complex between Ca^{2+} cations and a complexing agent. The complexing agent holds the calcium in solution, again preventing deposition of calcium salts or formation of scum. Typical complexing agents include **tripolyphosphates** and **polycarboxylates**. In these materials, negatively charged phosphate, $-PO_3^-$, or carboxylate groups, $-CO_2^-$, attract and hold the positively charged Ca^{2+} ions in solution, preventing the formation and deposition of $CaCO_3$ or scum.

3.4 REDOX CHEMISTRY

In this section we will be considering the processes of **oxidation** and **reduction**, and seeing examples of their application to various chemical species of importance in the environment.

3.4.1 OXIDATION, REDUCTION, AND OXIDATION STATES

Oxidation can be defined as either a gain in oxygen, for example, nitrite, NO_2^-, gaining an oxygen to become nitrate, NO_3^- (Equation 3.51), or a loss of electrons, as in Fe^{2+} losing one electron to become Fe^{3+} (Equation 3.52), or a loss of hydrogen, as when for example bright yellow deposits of elemental sulfur, S_8, are formed from the oxidation of hydrogen sulfide gas, H_2S, near volcanic vents or hot springs (Equation 3.53). In fact, more than one process usually occurs at the same time, as in the oxidation of metallic iron to Fe_2O_3 (ferric oxide or rust), in which each iron atom loses three electrons as well gaining oxygen atoms.

$$2NO_{2(aq)}^- + O_{2(g)} \longrightarrow 2NO_{3(aq)}^- \tag{3.51}$$

$$Fe^{2+} \longrightarrow Fe^{3+} + e^- \tag{3.52}$$

$$2H_2S_{(g)} + O_{2(g)} \longrightarrow 1/4S_{8(s)} + 2H_2O_{(l)} \tag{3.53}$$

The opposite of oxidation is **reduction**, which can be defined as a gain in electrons, a loss of oxygen, or a gain in hydrogen, which is the reverse of the three processes described earlier, as in the reduction of sulfate to hydrogen sulfide (Equation 3.54). You should note, however, that when an acid loses H^+ or a base gains H^+ as described in Section 3.2, this does not count as oxidation or reduction.

$$2H_{(aq)}^+ + 8e^- + SO_{4(aq)}^{2-} \longrightarrow H_2S_{(g)} + 4O^{2-} \tag{3.54}$$

Because oxidation and reduction involve the transfer of electrons, oxygen atoms, or hydrogen atoms, oxidation cannot take place without reduction, and reduction cannot occur without oxidation. The oxidation of one species is always accompanied by the reduction of another, and vice versa. A reaction in which oxidation and reduction take place is referred to as a **redox** reaction.

The extent to which a particular atom is oxidised or reduced can be indicated by its **oxidation state**. Atoms that have lost electrons or are combined with oxygen are **oxidised**, and have positive oxidation states. Atoms that have gained electrons or are combined with hydrogen are reduced, and have negative (or less positive) oxidation states. The oxidation state of a particular atom may be indicated by Roman numerals either as a superscript after the element symbol, for example, S^{VI}, which has an oxidation state of +6, or in brackets after the symbol, for example, Fe(III), which has an oxidation state of +3, or as a charge on an ion, for example, Zn^{2+}, which has an oxidation state of +2, and O^{2-}, which has an oxidation state of −2. Oxidation states are a means of chemical bookkeeping to ensure that all electrons have been accounted for. The oxidation state of an element in a molecule or ion can be determined by the following rules:

1. An element in its elemental state (the form it is in when it is not combined with other elements) has an oxidation state of 0. Examples include elements such as argon, Ar, or helium, He, which are monatomic gases; oxygen, O_2, nitrogen, N_2, or phosphorus, P_4, which form simple molecules; and iron, Fe, and copper, Cu, which are metals.
2. Ions that are formed from only one atom, such as Na^+, Cu^{2+}, Fe^{3+}, Cl^-, or S^{2-}, have an oxidation state that is simply the charge on the ion.
3. The oxidation state of hydrogen in compounds or ions is almost always +1, for example, in methane, CH_4, and water H_2O. (There is an exception to this rule when hydrogen is directly combined with a metal, but we will not consider this situation here.)
4. The oxidation state of oxygen in compounds or ions is almost always −2, as in ferric oxide, Fe_2O_3, or sulfate, SO_4^{2-}. (Again, we will not consider any of the exceptions here.)
5. The sum of oxidation states of all the atoms in a neutral molecule add up to 0, and the sum of oxidation states of all the atoms in an ion add up to the charge on the ion.

To show you how these rules work, we will look at two examples. Consider the molecule H_2SO_4, sulfuric acid. The molecule is a neutral molecule, so all the oxidation states add up to 0. The oxidation state of the two hydrogen atoms is +1 and that of the four oxygen atoms is −2, according to the preceding rules. We can now calculate the oxidation state of the sulfur atoms. Letting x be the oxidation state of the sulfur, we find that the oxidation state of sulfur in sulfate is +6:

$$0 = 2 \times (+1) + 4 \times (-2) + x$$

so,

$$0 = 2 + -8 + x$$

and therefore

$$x = +8 - 2 = +6$$

Consider now the oxidation state of nitrogen in nitrate, NO_3^-. The ion has a charge of −1, so all the oxidation states add up to −1. The oxidation state of the three oxygen atoms is −2, as before. Letting y be the oxidation state of the nitrogen atom, we find that the oxidation state of nitrogen in the nitrate ion is +5:

$$-1 = 3 \times (-2) + y$$

so

$$-1 = -6 + y$$

and therefore

$$y = -1 + 6 = +5$$

SELF-ASSESSMENT QUESTIONS

Q3.23 Give the oxidation state of every element in the following species:
  (i) H_2S
  (ii) NO_2^-
  (iii) NH_3
  (iv) Ca^{2+}
  (v) Fe_2O_3
Q3.24 The oxidation state of iron in iron pyrites, FeS_2, is +2. What is the oxidation state of the sulfur, S?
Q3.25 What is the oxidation state of sulfur in sulfate, SO_4^{2-}?

It is important that you realise that the formal oxidation state of an atom in a molecule or ion does not necessarily represent the actual charge on the atom. For example, although the oxygen atom in a water molecule has an oxidation state of -2, it does not actually carry a -2 charge. In fact, as you saw in Chapter 1, Section 1.4.4, the oxygen atom in water carries a slight negative charge, $\delta-$, because of its electronegativity.

3.4.2 REDOX POTENTIALS AND STABILITY FIELD DIAGRAMS

An environment can be described as reducing if it encourages the formation of reduced species, or oxidising if it encourages the formation of oxidised species. **Redox potential**, Eh, is a measure of how reducing or oxidising a particular system or environment is, i.e., its tendency to oxidise or reduce chemical species. Eh is related to the concentration of electrons in the system, and the more negative the value of Eh, the more reducing the environment. The redox potential of a particular system will determine the chemical form and solubility of those elements present that occur in different oxidation states, for example, whether iron is present as soluble Fe^{2+} or insoluble Fe(III) oxide, or whether sulfur is present as sulfide or sulfate. Environments in which O_2 is present are referred to as **oxic**, whereas environments in which O_2 is absent are referred to as **anoxic**. Oxic environments are oxidising, and have high Eh values, whereas anoxic ones are reducing and have low Eh values.

As we have already seen (for example, in the ionisation of phosphate and carbonate, or the bioavailability of Al^{3+}), the chemical form is also affected by pH. A **stability field diagram** shows what chemical species are likely to be present for a given element for different combinations of pH and Eh. A somewhat simplified

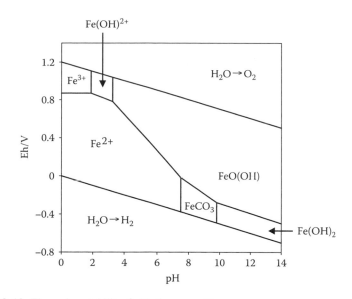

FIGURE 3.12 The redox stability field diagram of iron.

stability field diagram for iron is shown in Figure 3.12 as an example. The first point to note is that redox potential in an aqueous environment has upper and lower limits. The upper limit marks the potential above which water itself oxidises to form molecular oxygen, O_2, and the lower limit marks the potential below which water is reduced to form molecular hydrogen, H_2. Between these limits, different iron species are present depending on the conditions.

The diagram is dominated by two species. At low Eh values and acidic to just basic pH values, iron exists in the +2 oxidation state as the simple Fe^{2+} hexaaqua ion $[Fe(H_2O)_6]^{2+}$. At high Eh values and moderately acidic to strongly basic pH values, iron exists in the +3 oxidation state as insoluble hydrated ferric oxide, $FeO(OH)$, which is often (though not strictly correctly) written as $Fe(OH)_3$. Outside of the stability fields of these two species, other species such as iron(III) in the form of soluble $[Fe(H_2O)_6]^{3+}$ or $[Fe(H_2O)_5(OH)]^{2+}$, and iron(II) in the form of insoluble $FeCO_3$ and $Fe(OH)_2$ are formed.

A simplified stability field diagram for sulfur is given in Figure 3.13. This shows that sulfur occurs predominantly in its most oxidised form (sulfate), whereas sulfur in its lowest oxidation state of –2 only occurs in strongly reducing conditions. These are common in anoxic sediments where bacteria such as *Desulfovibrio desulfuricans* use dissolved sulfate as a source of oxygen to metabolise organic matter, forming H_2S in the process (Equation 3.55). At neutral to basic pH, sparingly soluble metal sulfides such as PbS, CdS, and ZnS, which all have very low solubility products, then readily precipitate out (Equation 3.56).

$$2CH_2O_{(s)} + SO_{4(aq)}^{2-} \longrightarrow 2HCO_{3(aq)}^{-} + H_2S_{(g)} \quad (3.55)$$

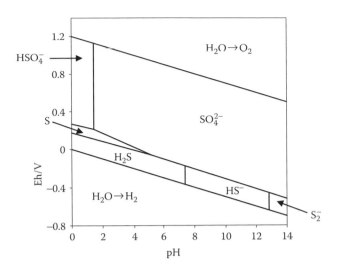

FIGURE 3.13 The redox stability field diagram of sulfur.

$$M^{2+}_{(aq)} + H_2S_{(aq)} \longrightarrow MS_{(s)} + 2H^+_{(aq)} \qquad (3.56)$$

One species not marked on either of the simplified iron or sulfur stability diagrams is iron pyrites, FeS_2. It is common in mineral veins, but it is also found in oil shales and coal, having been formed in anoxic conditions in organic-rich sediments from the reaction of initially formed FeS with dissolved H_2S (Equation 3.57).

$$FeS_{(s)} + H_2S_{(aq)} \longrightarrow FeS_{2(s)} + H_{2(g)} \qquad (3.57)$$

SELF-ASSESSMENT QUESTIONS

Q3.26 There are yellow deposits of sulfur in many volcanic regions. The gases emitted from vents in these areas are typically very acidic, and often contain H_2S, as indicated by a "rotten egg" smell. Can you explain why the sulfur deposits might form?

Q3.27 Approximately where in the sulfur and iron stability diagrams would you expect to see FeS_2?

In natural waters, conditions tend to be either strongly oxidising, (i.e., oxic) or strongly reducing (i.e., anoxic), so that we are usually concerned only with the upper or lower regions on stability field diagrams. The approximate positions of various environments are marked on Figure 3.14. Generally speaking, bodies of water are well-oxygenated and are therefore oxidising environments. An exception occurs when bodies of water are eutrophic (Section 3.6.2), in which water becomes anoxic. Bogs, waterlogged soils, and sediments, by contrast, are usually depleted in oxygen, often due to decaying organic matter that uses up oxygen in the decay process, and

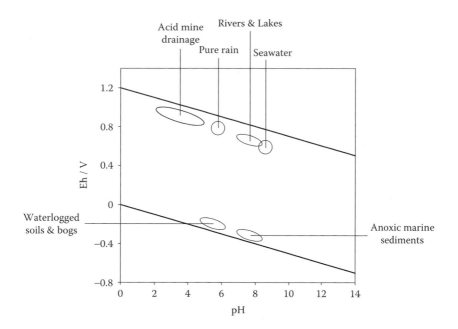

FIGURE 3.14 Stability field diagram for natural water courses.

these are therefore usually reducing environments. We will discuss the particular case of acid mine drainage in Box 3.4.

3.4.3 SPECIATION

Elements can exist in different forms in the environment. For example, sulfur can be in the solid form as elemental sulfur, S_8, metal sulfides, MS, metal sulfates, MSO_4, and iron pyrites, FeS_2, or in solution as the sulfate anion, SO_4^{2-}, or as a gas as hydrogen sulfide, H_2S, or sulfur dioxide, SO_2. Speciation is the term given to the existence of different physical or chemical forms of an element (especially in the natural environment). An important aspect of speciation is that the **toxicity** and **bioavailability** of an element can vary quite considerably depending on the form that that element is in. For example, chromium, Cr, in its most stable oxidation state of +3 as the Cr^{3+} ion, is less toxic than chromium in its most oxidised state of +6 as the **oxoanion** CrO_4^{2-}. Mercury, Hg, is generally toxic, but whereas mercury sulfide, HgS is insoluble and will be held in anoxic sediments in the aquatic environment, methyl mercury, CH_3Hg^+, which is formed biologically, is soluble and can get taken into the food chain. It is soluble in fat, and therefore tends to accumulate in fatty tissues.

The speciation of an element will depend on a number of factors such as pH; the presence or absence of complexing anions such as chloride, or the anions of sparingly soluble salts such as sulfide; the redox potential of the environment; and whether there are any biological mechanisms that can utilise the element. These factors will influence, for example, the solubility of the species, its oxidation state, and whether it is present as a neutral, positively charged, or negatively charged species. The speciation of an element will affect its bioavailabity. For example, nitrogen as N_2 can only be used by a small number of organisms, whereas nitrogen as ammonia or nitrate is readily taken up into plant roots (see the next section). One situation that can cause problems is where one species can mimic another in a chemical or biological environment with potentially serious results. An example of this is arsenate, AsO_4^{3-}, which is a negatively charged oxoanion that can mimic the similar but biologically active phosphate ion, PO_4^{3-}, and can replace it in chemical reactions in cells in living organisms with deleterious effects. Arsenic is in the same group of the periodic table as phosphorus, which is why its chemical behaviour is similar.

3.4.4 THE REDOX CHEMISTRY OF NITROGEN

Nitrogen is an essential nutrient for growth, and nitrogen-containing fertilizers are widely used in agriculture, but given that the atmosphere contains 78% by volume of nitrogen, why is it necessary to add extra nitrogen to soils at all? Nitrogen in its elemental state, N_2, is a very unreactive molecule because there is a strong triple bond between the two nitrogen atoms, and plants are unable to disrupt this bond. Only a handful of plant species, such as clover and peas are able to use atmospheric N_2 because it is **fixed**, i.e. converted to a useable form, by microorganisms that live in their roots. Plants take up nitrogen from the soil mainly in the form of fully oxidised nitrate, NO_3^-, or fully reduced ammonia, NH_3 or ammonium, NH_4^+, as these are the forms that occur in the redox conditions typical of water and soils, as shown in the nitrogen stability field diagram in Figure 3.15. There are naturally occurring sequences of reactions in which nitrate is reduced and ammonia or ammonium is oxidised, and you will read more about these processes in Chapter 4, Section 4.3.2.

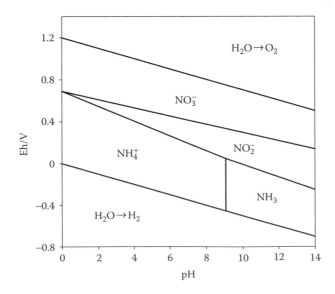

FIGURE 3.15 The stability field diagram for nitrogen.

TABLE 3.5
Oxidation States of Nitrogen

Oxidation State	Example	Formula
+5	Nitrate	NO_3^-
+4	Nitrogen dioxide	NO_2
+3	Nitrite	NO_2^-
+2	Nitric oxide	NO
+1	Nitrous oxide	N_2O
0	Nitrogen	N_2
−1	Hydroxylamine	NH_2OH
−2	Hydrazine	N_2H_4
−3	Ammonia, ammonium	NH_3, NH_4^+

The available oxidation states of nitrogen range from −3 to +5, and are given in Table 3.5 together with examples of compounds that occur in the sequences of oxidation and reduction reactions involving nitrogen. **Nitrification** is the process in which ammonia or ammonium is oxidised to nitrite, NO_2^-, and then to nitrate via hydroxylamine, NH_2OH, whereas **denitrification** is the reduction of nitrate to molecular nitrogen, N_2, or sometimes nitrous oxide, N_2O, via nitrite and nitric oxide, NO. The two general sequences of reactions are shown in Equations 3.58 and 3.59. Many of the reactions are mediated by bacteria, with nitrification generally taking place in oxic conditions and denitrification generally taking place in anoxic conditions. Nitrifying bacteria (e.g., *Nitrosomonas*) and denitrifying bacteria (e.g., *Paracoccus denitrificans*) are used to treat waste water to remove excessive levels of nitrate or ammonia.

$$NH_3 \longrightarrow N_2H_4 \longrightarrow NH_2OH \longrightarrow NO_2^- \longrightarrow NO_3^- \qquad (3.58)$$

$$NO_3^- \longrightarrow NO_2^- \longrightarrow NO \longrightarrow N_2O \longrightarrow N_2 \qquad (3.59)$$

Box 3.4 Acid Mine Drainage

A pollution problem that occurs as a result of mining is acidification of water courses. This is referred to as **acid mine drainage**. When rock is exposed to air as a result of mining, any sulfide minerals that are present (either as exposed mineral veins or as waste deposits) will be oxidised to soluble sulfate, forming H^+ ions at the same time, thus generating sulfuric acid, as in the oxidation of iron pyrites:

$$FeS_{2(s)} + 7/2 O_{2(g)} + H_2O_{(l)} \longrightarrow Fe^{2+}_{(aq)} + 2SO_4^{2-}_{(aq)} + 2H^+_{(aq)}$$

At the very acidic pHs typical of acid mine drainage, Fe^{2+} is fairly stable and oxidation to iron(III) is slow, but the pH of the water will increase with increasing distance from the mine, and then Fe^{2+} will readily oxidise to iron(III). This will precipitate as an oxide in the form of hydrated ferric oxide, Fe_2O_3, or goethite, $FeO(OH)$, forming extensive red-coloured deposits, and causing a reddish discoloration of the water due to colloidal particles of the oxide.

$$4Fe^{2+}_{(aq)} + O_{2(g)} + 6H_2O_{(l)} \longrightarrow 4FeO(OH)_{(s)} + 8H^+_{(aq)}$$

The resulting highly acidic water is very damaging to the environment (see Box 3.2). The problem of acid mine drainage is exacerbated once mines are disused and are no longer pumped dry because water accumulates, increasing the amount of dissolved sulfuric acid. Water may then escape from the mine into the environment in an uncontrolled manner.

One potential solution to the problem of excess levels of iron in acid mine drainage currently being investigated is to treat it with a solution of phosphate, PO_4^{3-}. Insoluble iron(III) phosphate, $FePO_4$, will precipitate out, which can then be removed and reused as a fertilizer.

3.5 COLLOIDS AND SUSPENDED PARTICLES

We have considered how soluble ionic and polar molecules dissolve in water. We will now consider **colloids**, which are insoluble microscopic particles in water. There is an enormous density of colloids in natural waters, with more than 10^6 particles per cubic centimetre in fresh surface waters, groundwaters, oceans, and sediment waters. The conventional definition of a colloid is based on whether a particle will pass through a membrane: dissolved species (i.e., ions or polar molecules with some definite interaction with water as described earlier) will pass through, whereas larger particles will not. Colloidal particles are generally between 0.001 μm (1 nm) and 100 μm (0.1 mm) in size, but they have at least one dimension in the range 0.001–1 μm. Colloids are dynamic particles: they are continuously losing material and having new material added to them, precipitating and dissolving. Small colloids may aggregate, forming larger particles that can then settle out of solution. Colloids with particle sizes of less than about 10 μm do not settle out of solution. Sources of colloids include weathered rock, which forms SiO_2 (colloidal silica or H_4SiO_4), $Al(OH)_3$, $Fe_2O_3.nH_2O$ and clays; humus from soils; and biological debris such as dead plankton and faecal matter.

An important aspect of colloids is that they regulate concentrations of many elements, including heavy metals, and organic pollutants in water. They do this because ions and molecules adsorb to the surface of colloidal particles; the larger the available surface area for adsorption, the greater the number of ions and molecules that can be adsorbed. One estimate of the surface area of colloidal particles available for adsorption in the upper layers of seawater is that there is greater than

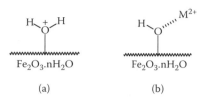

FIGURE 3.16 Adsorption of metal ions to hydrated ferric oxide: (a) surface protonated at low pH; (b) surface binding metal ions at higher pH.

about 18 m^2 per cubic metre of water. The surfaces of colloidal particles are charged and therefore hold ions and attract polar molecules. Heavy metals tend to adsorb onto the broken edges of inorganic crystals, where there are broken bonds that are often protonated. An example is the surface of hydrated ferric oxide, $Fe_2O_3.nH_2O$ (which can also be written as FeOOH). At low pH these hydroxyl groups are protonated and the metal ions cannot bind, but at higher pH values, metal cations can displace H^+, and thus bind to its surface (Figure 3.16). In a solution containing Cd^{2+} and colloidal $Fe_2O_3.nH_2O$, for example, the proportion of bound Cd^{2+} rises from 0% at pH 5 to 100% at pH 7. By contrast, anions tend to desorb at higher pH values; for example, the proportion of bound arsenic in the form of arsenate, AsO_4^{3-}, drops from 100% at pH 9 to 0% at pH 13. You will learn in Section 3.6.1 that the ability of iron(III) to bind arsenate and the liberation of arsenate when iron(III) is reduced to iron(II) has profound consequences for water supplies in several parts of the world.

3.6 WATER POLLUTION

Water can become polluted through direct input to water courses, e.g., effluent from factories and sewage works, or via runoff from land into surface waters, or through contamination of groundwater. We will discuss a range of pollutants, grouping them generally into heavy metals, nutrients, and organic pollutants.

3.6.1 HEAVY METALS

Heavy metal pollution of waters is, as with soil pollution, of serious concern. Metals such as mercury, cadmium, lead, chromium, etc., are all toxic and pose problems. We will mention one recent historical case and two issues of serious current concern to highlight examples of how heavy metal pollution can occur. Mercury is used in the manufacture of a wide range of industrial products such as thermometers, lamps, and electrical apparatus. Inorganic mercury species such as the Hg^{2+} ion can be converted to organic mercury compounds such as methyl mercury, CH_3Hg^+, over time, and these in particular accumulate in aquatic and marine organisms because they are soluble in fatty tissues. A well known example of mercury poisoning occurred in Minamata, Japan. For many years mercury was discharged into the sea near Minamata by local industries. The mercury accumulated in fish, which was subsequently eaten by local people. Over a 20-year span up to 1975, more than 100 people died and many others suffered severe health effects from eating the contaminated fish.

Tin pollution of coastal waters is occurring in particular as a result of the use of tributyltin compounds, $(C_4H_9)_3SnX$ (where X is chlorine, fluorine, hydroxide, etc.), in marine antifouling paints. These are used on the hulls of ships and boats to prevent or reduce the growth of barnacles, algae, etc., which will increase the drag on the hull and slow a boat down. As the antifouling gets worn away or washed off a boat's hull, tin is leached into solution and subsequently accumulates in marine and aquatic organisms. Tributyltin-based antifouling paints are banned in the U.K. on vessels under 25 m in length.

A particular issue that has come to the fore in the last 20 years is the problem of arsenic-contaminated tube wells in parts of India, Bangladesh, and Latin America, among other countries. The tube wells were originally sunk to provide clean water for communities that were using untreated surface water for drinking, and suffering gastrointestinal diseases as a result. It has been discovered in the last few years, however, that the water in many wells is heavily contaminated with arsenic. The World Health Organisation (WHO) recommended maximum concentration of arsenic is 10 μg L^{-1} (10 ppb), and the Bangladesh standard is 50 μg L^{-1} (50 ppb). A recent survey found that 35% of tubewells have arsenic levels in excess of 50 ppb, whereas 8% have levels in excess of 300 ppb, affecting tens of millions of people. People who drink the contaminated water may suffer severe skin lesions (a symptom of chronic arsenic poisoning), and it is feared that many will develop arsenic-related cancers. The arsenic probably originates from buried sediments. It has been suggested that arsenic in the form of arsenate, AsO_4^{3-}, is held on the surface of FeOOH particles in these sediments. At some point after the sediments have been buried, Fe(III) is reduced to Fe(II), possibly by buried organic matter, thus releasing arsenate from the surface of the iron particles into the groundwater, which can then contaminate wells.

In order to be able to set and enforce limits on the concentrations of metals in drinking water, it is necessary to be able to determine the actual concentrations of the various metal species. The principal analytical technique for determining concentrations of metals in aqueous solutions is inductively coupled plasma optical emission spectroscopy (ICPOES or simply ICP), which is described in Box 3.5.

3.6.2 NUTRIENTS AND EUTROPHICATION

The base of the food chain in water is formed by phytoplankton, which are single-celled plants or algae that grow by **photosynthesis**. These are used as food by zooplankton (microscopic animals) and other creatures further up the food chain. Phytoplankton need nutrients to grow, including nitrate, NO_3^-, and phosphate, PO_4^{3-} (which is actually mostly as HPO_4^{2-} at the pH of seawater), but an excess of these nutrients leads to excessive growth of algae, which results in **eutrophication**. Two particular problems are created by excessive algal growth. The first is that thick blankets of algae at the surface of the water block light to plants that live further down the water column, causing a reduction in photosynthesis or even plant death and, hence, reducing the formation of oxygen. The second problem is that when algae die, oxygen is used up in the decay process. An excessive number of decaying algal cells can cause oxygen levels to become very depleted, particularly near the bottom of rivers, lakes, and shallow seas, where the decaying cells accumulate. This is termed eutrophication. Depleted oxygen levels result in the death of other aquatic life, and can lead to the "death" of rivers or lakes (i.e., the loss of aquatic life), or "dead zones" in some coastal waters.

High nitrate levels in rivers and groundwater is mainly due to the use of nitrate fertilizers in agriculture. Nitrate gets washed out of soil directly into rivers via surface runoff from land, but nitrate also gets carried down into groundwater. This poses a potential problem because groundwater is a major source of drinking water, and high nitrate levels are harmful to human health. Water with a high nitrate concentration that is intended for drinking may have to be diluted with water having a low concentration of nitrate in order to lower the nitrate concentration enough to meet national limits. The maximum permitted limit is 50 mg L^{-1} in the EU, as set by the Nitrates Directive of the European Union. There are a number of sources of phosphate, including phosphate fertilizers, animal manure, and detergents. Some animal feed has phosphate added to it, which increases the burden of phosphate in animal manure. Reductions in fertilizer use will lead to a reduction in surface runoff into water courses, but nutrients that have got into groundwater supplies will continue to cause problems for many years.

3.6.3 ORGANIC POLLUTANTS

The range of organic pollutants is large, but we will mention a few classes that are of concern in the aquatic environment. As with nutrient pollution, agricultural runoff is also a source of pesticides, herbicides, etc., which then get into surface and groundwater. These not only affect local aquatic life, but also pose problems for water companies that extract water for human consumption. There are strict maximum admissible levels on many pesticides, fungicides, herbicides, etc. There are also limits for other organic pollutants already mentioned in Chapter 2, Section 2.7.3 such as PCBs and PAHs, for example, benzene and benzo[a]pyrene. Some hazardous chlorinated compounds are formed following chlorination treatment of water, for example, 4-chloro-3-methylphenol, or as a result of reactions between organic compounds and chlorine-containing cleaning products.

A class of organic pollutants causing especial concern is **endocrine disruptors**. These are molecules that affect sexual development and fertility, and one effect they are having in the environment is in the feminisation of male fish. This in turn affects the ability of male fish to breed and could potentially have a disastrous effect on freshwater and estuarine fish populations. The principal source of endocrine disruptors is effluent from sewage treatment works, which contains the natural steroid hormones oestrogen and oestradiol, and a related synthetic variant. These are excreted by women taking the contraceptive pill. A minor (in comparison to the steroid hormones) class of endocrine disruptors are **phthalates,** which are used mostly in plastics to provide flexibility, especially to PVC (polyvinylchloride). These are now ubiquitous in the environment as they leach readily from plastic waste, as well being discharged in waste or emissions from factories.

3.6.4 MARINE OIL POLLUTION

The final class of organic-based water pollution that we will consider is marine oil spillage. Spillages occur on land as well, but the amounts of oil spilled, the environmental problems caused and the difficulties in employing countermeasures make marine oil spills particularly problematic. There have been a number of high-profile major spillages such as the *Torrey Canyon*, (Land's End, U.K., 1967), the *Amoco Cadiz* (Brittany, France, 1977), the *Exxon Valdez* (Prince William Sound, Alaska, 1989), and the *Sea Empress* (Milford Haven, U.K., 1996) because of oil tankers running aground. The main environmental concerns with marine spillages are the threat to large numbers of sea birds and damage to fragile coastal ecosystems such as salt marshes, mud flats, and shell fish beds. There are a number of measures that can be taken in the event of a marine spillage such as mechanical recovery of the oil, burning off, and use of dispersants. Volatile components evaporate readily (the nature of the oil that has been spilled determines how much is lost this way) and, sometimes, rough weather causes slicks to break up naturally. Wind speeds above Force 3 (above about 16–17 km hr^{-1}) and wave heights in excess of 1–2 m make mechanical recovery very difficult.

A very important chemical means of treatment (used with good success in the *Sea Empress* spillage) is the use of dispersants. Dispersants are a class of chemical called **surfacants** (surface-active agents). The principles on which they work are exactly the same as in domestic detergents such as washing-up liquid, and indeed some of the surfactants used to treat oil spills occur in household products such as shampoos. Surfactants consist of an ionic or polar head and a long nonpolar hydrocarbon chain. Figure 3.17 shows an example with an ionic head, the zigzag representing the long hydrocarbon chain. The surfactant molecules form a **micelle**: the nonpolar, hydrophobic tails dissolve in a droplet of oil, whereas the polar or ionic hydrophilic heads point outwards and hold the droplet in the surrounding water. This causes the droplet to break off from the slick and get dispersed into the water column. The overall effect is to enhance the concentration of oil in the water column, but this is really only a problem in shallow water. Dispersants are generally best used early, before an emulsion called a **mousse** forms. This is a water-in-oil emulsion; i.e., droplets of water are dispersed in the oil, which is then very difficult to disperse.

Box 3.5 Analysis of Metals in Aqueous Solutions

In Box 2.1 we described how metal concentrations in soils can be described. We now describe how the levels of metals in aqueous solutions can be achieved. The most widely used technique is inductively coupled plasma optical emission spectroscopy (ICPOES, or simply ICP), which is a form of atomic spectroscopy. In ICP, a sample of the water is **aspirated** (sucked up) into the ICP spectrometer, where the sample passes into an argon plasma (i.e., a mix of argon ions and electrons in the gaseous state) that is at a temperature of about 10000 K. At this temperature the sample is atomised (converted into atoms in the gas phase) and ionised (loses electrons to form cations). The high temperatures cause electrons

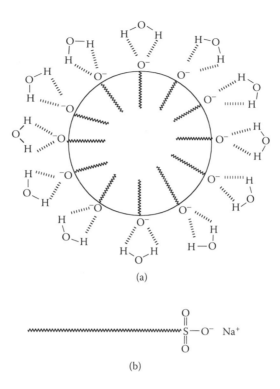

(a)

(b)

FIGURE 3.17 Dispersants in treatment of oil spills: (a) showing formation of a micelle, with nonpolar tails in the oil droplet and charged group in the surrounding water; (b) detail of a typical dispersant showing the charged hydrophilic head, and the long nonpolar (hydrophobic) hydrocarbon chain.

in the atoms and ions to jump to higher energy levels, and when the electrons drop back they emit radiation at visible and UV wavelengths. Each element emits radiation of specific wavelengths. The emitted radiation is split into its component wavelengths and lands on a detector (usually a silicon chip) that detects the wavelengths and intensities of all emitted light simultaneously. As with XRF analysis, the identities of the different elements present are determined by the different wavelengths of light being emitted, and the amount of each element is determined by the intensity of the emitted radiation.

An alternative form of atomic spectroscopy widely used to determine low levels of heavy metals such as mercury is graphite furnace atomic absorption spectroscopy (GF-AAS). In this technique, the sample is placed in a small graphite tube (about 3 cm long). The tube is heated electrically to about 3000 K, which atomises the sample. A beam of light whose wavelength is specific to the element being determined is passed along the graphite tube. Some of the light is absorbed by the atoms of the element of interest, and the amount of light absorbed is directly proportional to the amount of that metal present in the sample.

3.7 SUMMARY

In this chapter you should have learned that:

Water is a remarkable compound, and many of its properties are due to its ability to hydrogen-bond. One of its most important properties is the anomalous expansion of water. Water is also responsible for transporting huge quantities of heat around the globe.

An acid is a substance that can donate one or more protons. A base is a substance that can accept one or more protons. An acid and its conjugate base form a conjugate acid–base pair.

Acids are classed as strong or weak, depending on whether they are fully or partially ionised in water. Weak acids have an acid dissociation constant, K_a, which quantifies the extent of dissociation.

pH is a measure of the acidity of a system. pH is a logarithmic scale in which a pH of 7 is neutral, pH values less than 7 are acidic, and pH values greater than 7 are basic.

The concentration of acid or base in a solution can be determined by an acid–base titration.

A buffer solution hardly changes pH when small volumes of acid or base are added to it.

Ions in solution have a hydration sphere of water molecules around them. Some metal cations form specific complexes with water or other ligands such as chloride.

In the carbonate system there is a series of equilibria involving $CO_{2(g)}$, $CO_{2(aq)}$, $H_2CO_{3(aq)}$, $HCO_3^-{}_{(aq)}$, and $CO_3^{2-}{}_{(aq)}$. $CaCO_3$ is a sparingly soluble salt.

Oxidation states are a form of chemical bookkeeping.

Stability field diagrams show what chemical forms an element is in at different pH and Eh levels.

The speciation (the different chemical forms) of an element in the environment is an important factor in the bioavailability of that element and its potential beneficial or harmful effects to organisms.

Now try the following questions to test your understanding of the material covered in this chapter.

Q3.28 (i)   What physical state (i.e., solid, liquid, or gas) is water in at 0.5 atm and 0°C?

(ii) What happens if the pressure is increased while maintaining constant temperature?

(iii) What happens if the temperature is increased while maintaining constant pressure?

Q3.29 An ocean current has a flow of water of 60×10^6 m^3 s^{-1}, and the temperature drop along the flow is 4°C. How much heat is being transported?

Q3.30 In the following equation, identify the acid, the base, the conjugate acid, and the conjugate base.

$$Na_2CO_3 + HCl \longrightarrow 2Na^+ + Cl^- + HCO_3^-$$

Q3.31 (i)   The pH of a sample of mineral water is pH 8.5. Calculate $[H^+]$.

(ii)   The concentration of HCO_3^- in this sample is 136 mg L^{-1}. Convert this to a concentration in mol L^{-1}.

(iii) K_{2a} for HCO_3^- is 4.68×10^{-11} mol L^{-1}. Calculate $[CO_3^{2-}]$ in units of mol L^{-1}.

(iv)   Calculate the alkalinity of this sample in units of mol L^{-1}.

(v)   Calculate the alkalinity of this sample in units of mg L^{-1} $CaCO_3$.

Q3.32 (i) K_{sp} for $CaCO_3$ is 3.8×10^{-9} mol^2 L^{-2}. Calculate the concentration of Ca^{2+} at which precipitation of $CaCO_3$ is expected to occur if $[CO_3^{2-}] = 2.45 \times 10^{-5}$ mol L^{-1}

(ii)   Convert the concentration of Ca^{2+} in mol L^{-1} to a concentration in mg L^{-1}.

Q3.33   The concentrations of Ca^{2+} and Mg^{2+} in a sample of mineral water are 28.5 mg L^{-1} and 2.3 mg L^{-1}, respectively. Calculate the hardness of this water sample.

Q3.34   Give the oxidation states of all the species in $NaHCO_3$.

4 Air

In this chapter we will consider the structure and composition of Earth's atmosphere, as well as two major issues concerned with the atmosphere: global warming and ozone depletion. Other aspects of air pollution will also be considered, together with the cycling of carbon, nitrogen, and sulfur through the environment.

4.1 THE STRUCTURE OF THE ATMOSPHERE

The vertical structure of the atmosphere is given in Figure 4.1. The **troposphere** is 8–15 km thick, the lower value being at the poles, and it is subject to seasonal variation. The troposphere is where most weather systems develop. Above the troposphere is the **stratosphere**, where the air flow is less turbulent and has fewer clouds. The boundary between the two is called the **tropopause** and is marked by the height at which temperature stops decreasing with height and begins to increase with height. Air exchange between the troposphere and stratosphere is slow, and therefore air pollution (which we will be considering in Section 4.6) is generally contained with the troposphere, but if pollutants are injected into the stratosphere they may remain there a long time. Examples of naturally occurring species that get injected into the stratosphere are dust and aerosols from volcanic eruptions, which can stay in the atmosphere for years. The stratosphere extends up to about 55 km. Above the stratosphere is the mesosphere, which extends to about 85 km and above that there is the thermosphere. The thermosphere is the region where auroras occur —the Aurora Borealis (Northern Lights) in the northern hemisphere and the Aurora Australis (Southern Lights) in the southern hemisphere. You can read about auroras in Box 4.1.

The main driving force for the circulation of the atmosphere is **incident solar radiation**. Air in the equatorial regions is heated, rises, and moves poleward before sinking and returning to the equator. Rotation of the Earth forces this circulation into a spiral formation. Roughness (e.g., buildings, vegetation, etc.) and contours on the ground have only a small effect on the overall pattern of atmospheric circulation, and at most altitudes air movements approximate to a nonviscous (i.e., free-flowing) fluid. The theoretical wind speed, generated by the rotation of the Earth is termed the **geostrophic wind speed**, but it is also affected by changes in pressure. Near to the ground is a boundary layer, about 1 km thick, where friction (from land contours, buildings, trees, etc.) and buoyancy (i.e., rising air that has been heated by hot ground) generate turbulence. Over rough ground the geostrophic wind speed may drop by 40%, whereas over the sea the drop may be around 20%. Frictional effects also result in a variation of wind direction with height, which is called **wind shear**.

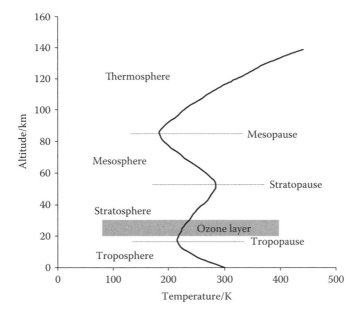

FIGURE 4.1 The structure of the atmosphere.

All these effects may impact on the nature, location, and severity of air pollution (Section 4.6).

Box 4.1 The Formation of Auroras

About 100 km above the Earth, energetic charged particles (mostly protons and electrons) emitted by the Sun are trapped as they encounter the Earth's magnetic field. From time to time, and especially when the Sun is more active than usual, the trap fills up and overflows, spilling fast-moving protons and electrons into the upper atmosphere, at high latitudes close to the Earth's magnetic poles. As these particles crash into the atmosphere, they knock electrons off the atoms and molecules they find there, ionizing them. Subsequently, the electrons recombine with the ionized atoms and molecules. The re-formed atoms and molecules have high energies and can lose some of their excess energy by giving out light.

The colour of the light depends on which atoms or molecules are ionized. The commonest auroral colour is whitish green, which is emitted by high-energy (**excited**) oxygen atoms, O^*. When ionised molecular oxygen recombines with an electron, the resulting O_2 molecule is very energetic and splits up into two O atoms (shown below), some of which are in a high-energy state, O^*. O^* emits light, giving out energy in the process. Most O^* emit whitish green light with a wavelength of 557.7 nm, whereas a few emit reddish light with a wavelength of 630 nm. (The wavelength range of visible light extends from the blue, short wavelength end at 400 nm to the red, long wavelength end at 750 nm.)

$$O_2 + \text{charged particle} \longrightarrow O_2^+ + e^-$$

$$O_2^+ + e^- \longrightarrow O* + O$$

$$O* \longrightarrow O + \text{light}$$

Violet light with a wavelength of 427.8 nm is emitted by ionised molecular nitrogen, N_2^+. OH· **radicals** (chemical species with an unpaired electron), formed by the reaction of O atoms with H_2O, emit red light called **airglow** at 730 nm. Other colours are very rarely seen.

4.2 EVOLUTION AND COMPOSITION OF THE ATMOSPHERE

The chemical composition of Earth's atmosphere is shown in Table 4.1. It consists primarily of nitrogen and oxygen with very small amounts of other gases including

TABLE 4.1
The Composition of the Atmosphere

Gas	Formula	Concentration (by volume)
Nitrogen	N_2	78.08%
Oxygen	O_2	20.95%
Water[a]	H_2O	0.5-4%
Argon	Ar	0.934%
Carbon dioxide[b]	CO_2	0.0382%, 382 ppmv
Neon	Ne	18 ppmv
Helium	He	5.2 ppmv
Methane[a]	CH_4	2.0 ppmv
Krypton	Kr	1.1 ppmv
Hydrogen	H_2	0.5 ppmv
Nitrous oxide	N_2O	0.3 ppmv
Carbon monoxide[a]	CO	0.1 ppmv
Xenon	Xe	0.087 ppmv
Ozone[a]	O_3	0.045 ppmv
Ammonia[a]	NH_3	0.006 ppmv
Nitrogen dioxide[a]	NO_2	0.001 ppmv
Sulfur dioxide[a]	SO_2	0.0002 ppmv
Hydrogen sulfide[a]	H_2S	0.0002 ppmv

[a] Variable concentration; water vapour can vary from 0.5 to 4%.
[b] Concentration increasing by 1–2 ppmv per year.

water, argon and carbon dioxide. The concentration of water vapour is rather variable, whereas the concentration of carbon dioxide is increasing by about 1–2 ppmv (parts per million by volume) per year largely as a consequence of the burning of fossil fuels. There are also trace quantities of other gases, including ozone, O_3, which will be discussed in Section 4.5. This atmospheric composition is very different from that originally formed during our planet's evolution. Earth's **primary** atmosphere, which derived from the cloud of gas and dust from which the Earth formed, consisted largely of hydrogen and helium, but was blown away very quickly by the solar wind. By analysing the ratio of ^{40}Ar and ^{129}Xe (Chapter 1, Section 1.1.6), it has been determined that about 85% of the Earth's **secondary** atmosphere was released from the mantle (outgassed) very early in Earth's history and the remainder has slowly been released through volcanic activity. Volcanic gases contain very much more CO_2 than N_2 and, ignoring the water content (the water content in terrestrial volcanoes is derived from surface water and not the mantle), this composition also closely resembles the atmospheres of Mars and Venus. Thus, we can conclude that Earth's early secondary atmosphere contained very large amounts of CO_2. This leads us to the question, why is the Earth's present atmosphere so different?

As the Earth cooled after its formation, the surface temperature dropped low enough for liquid water to form, allowing CO_2 to dissolve. The high levels of atmospheric CO_2 would have meant that the oceans would have been able to dissolve more CO_2 (Henry's Law; Chapter 3, Section 3.3.3). Reaction of basalts erupted on the seabed with dissolved CO_2, as well as weathering of the small amount of continental rocks (Chapter 2, Section 2.6.2) would have led to the deposition of large amounts of carbonate rock, and this may have accounted for a significant reduction in the level of CO_2 over the first 1 billion or so years of Earth's existence.

Life in the form of simple bacteria had evolved by 3.5 billion years ago, and this resulted in CO_2 being converted into organic carbon compounds. From this point on, burial of organic matter was another sink for atmospheric CO_2, including, for example, the vast forests of Carboniferous times that formed many of the Earth's coal deposits, and the accumulated remains of plankton that formed oil and natural gas (Box 2.3). As we burn fossil fuels, this carbon is being re-released into the atmosphere as CO_2 at a much faster rate than would otherwise occur. The biological deposition of $CaCO_3$ as shells and tests of marine organisms, as described in Chapter 2, Section 2.5.3, is yet another factor in reducing carbon dioxide levels in the atmosphere to the current levels of a few hundred ppmv.

Although CO_2 levels have fallen dramatically, the concentration of atmospheric oxygen has increased equally dramatically. Figure 4.2 shows how the level of oxygen has increased over Earth's history. In the early Earth, virtually all oxygen was in a combined form as CO_2 and H_2O, and there was very little free molecular oxygen, O_2. Two ways of producing O_2 are **photolytic dissociation** and **photosynthesis**. Photolytic dissociation occurs when ultraviolet light interacts with water molecules high up in the Earth's atmosphere, breaking the H-O bond and releasing hydrogen and oxygen atoms. Once dissociated, the hydrogen and oxygen atoms can either recombine to form H_2O, or combine with other hydrogen and oxygen atoms to form H_2 or O_2. As H_2 is very light, it tends to

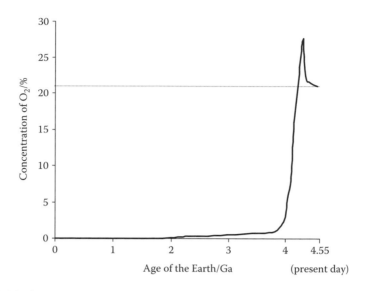

FIGURE 4.2 Concentration of atmospheric oxygen over Earth's history.

drift from the upper atmosphere into space, leaving the heavier O_2 molecule behind. This process leads to a reduction of water in our atmosphere, and up to about 1/3 of the water originally on Earth may have been lost this way. Photolytic dissociation of CO_2 may also have contributed to the formation of small amounts of O_2 in the atmosphere.

Although a small amount of O_2 is produced by photolytic dissociation, most O_2 is produced biochemically by photosynthesis. All green plants contain a number of pigment molecules, the most abundant of which are **chlorophyll a** and **chlorophyll b**. Both these pigments are based on a complex organic structure with a magnesium atom at the centre (Figure 4.3). These molecules absorb visible light, converting it to chemical energy, which is then used to convert CO_2 and water into plant matter such as cellulose, which has the general formula $(CH_2O)_n$ (Equation 4.1).

$$nCO_{2(g)} + nH_2O_{(l)} + sunlight \longrightarrow (CH_2O)_{n(s)} + nO_{2(g)} \qquad (4.1)$$

Initially, all oxygen formed in the atmosphere was used up in oxidising dissolved iron(II) in the oceans (mainly derived from underwater volcanic vents) to insoluble iron(III), as well as the oxidation of iron(II) to iron(III) in the weathering of exposed rocks. The first rocks containing fully oxidised red Fe_2O_3 have been dated to about 2.2 billion years ago (i.e., when the Earth was about 2.3 Ga old), indicating that atmospheric O_2 levels had risen to about 1–2%, and by 0.6 billion years ago (i.e., 3.95 Ga) virtually all the available dissolved iron(II) had been oxidised to iron(III). At this point free O_2 began to accumulate in the atmosphere. By 0.5 billion years ago (i.e., 4.05 Ga) the oxygen level had reached at least 6%. The maximum level of O_2 occurred in the Carboniferous, where levels may have reached as high as 35%. Today the atmospheric level of O_2 has stabilised at about 21%.

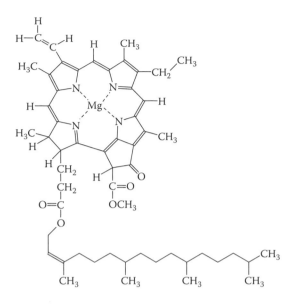

FIGURE 4.3 The structure of chlorophyll a.

SELF-ASSESSMENT QUESTION

Q4.1   Why do you think atmospheric O_2 levels did not exceed about 35%, and have stabilised at lower values?

Over the last few decades there has been much debate about the source of the organic molecules that must have been present on Earth for life to begin. Many meteors are known to contain simple organic molecules, and it has been suggested that life was 'seeded' from space, a theory known as **panspermia**. Another source of the required organic compounds that has been proposed is the atmosphere, and in particular, the products obtained from reactions of atmospheric gases. You can read more about this in Box 4.2.

Box 4.2 Formation of the Molecules of Life

Evidence in the form of fossil stromatolites (structures formed by mats of primitive bacteria and layers of sediment) indicate that life had started on Earth by 3500 million years ago, and variations in ^{12}C and ^{13}C isotope levels in organic carbon deposits suggest that life may have started even earlier than that. Then, how did life start? We do not know the answer to this, but another question that we can ask is, how did the molecules form that are essential to life? In the early 1950s, two American scientists, H.C. Urey and S.L. Miller passed an electrical spark (to mimic lightning) through a flask containing water vapour (H_2O), methane (CH_4), ammonia (NH_3), hydrogen sulfide (H_2S), and a small amount of hydrogen (H_2) representing Earth's primitive atmosphere. The resulting mix of organic compounds included several of the amino acids that are the building blocks of proteins, as well as three of the nitrogen-containing bases that are essential components of DNA and RNA, the molecules that carry the genetic code of all organisms. It seemed as if an answer had been found to the question of the source of the molecules of life. However, as you have read in Section 4.2, Earth's primitive atmosphere was probably mostly carbon dioxide (CO_2) and water vapour with some nitrogen (N_2), a very different composition from that used in Miller and Urey's experiments. Electrical spark experiments on combinations of gases similar to this have failed to give the range and yield of products found in Miller and Urey's experiments. A major difference is that Miller and Urey's gas mixture contained elements in **reduced** (combined with hydrogen) form, i.e., CH_4 and NH_3 instead of CO_2 and N_2, which is nearer to the form of these elements in amino acids and DNA.

Any mechanism for the **abiotic** (nonbiological) synthesis of the molecules of life must involve the **reduction** of carbon and nitrogen. Among current theories are suggestions that molecules were formed in the primitive ocean by reduction with Fe(II), either in solution as soluble Fe^{2+} or on the surface of solid FeS or $Fe(OH)_2$, or on the surface of clay particles in which the clay acts as a catalyst and as a means of holding simple molecules in close proximity, or on the surfaces of other particles such as metal oxides or sulfides. Research in this field continues.

4.3 BIOGEOCHEMICAL CYCLES

Carbon and other elements essential to life on Earth such as nitrogen and sulfur are cycled through the atmosphere, the **hydrosphere** (oceans, lakes, rivers, etc.), the **geosphere** (sediments and rocks), and the **biosphere** (living organisms and their ecosystems) in what are called **biogeochemical** cycles. Many of the stages of the various cycles occur naturally, but some are **anthropogenic** (due to human activity), and these can have significant impacts on the environment. In this section we will examine the biogeochemical cycles of carbon, nitrogen, and sulfur.

4.3.1 THE CARBON CYCLE

The consumption of CO_2 by plants described in the previous section is part of a much larger environmental cycle called the **carbon cycle**. A simplified version is shown in Figure 4.4, which shows the main **reservoirs** of carbon and the short term **fluxes** (movement of carbon). As you can see, carbon in biomass and soil forms only a small fraction of total carbon, whereas the oceans are the largest reservoirs for readily accessible carbon, with most present as hydrogen carbonate, HCO_3^-. However, this is dwarfed by the amount of carbon contained in sedimentary rocks occurring as calcium carbonate (limestone) and as buried organic matter, which together account for over 99% of the carbon in the Earth's crust.

The carbon cycle has different **timescales** for the fluxes between the various reservoirs, and the **residence time** of carbon (the average length of time spent by a

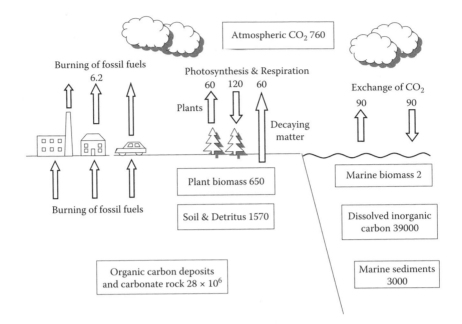

FIGURE 4.4 The carbon cycle, showing (a) the main reservoirs of carbon in 10^{12} kg C and (b) the major short-term fluxes of CO_2 in 10^{12} kg C yr^{-1}. (From A. Colling (ed.), *The Dynamic Earth*, The Open University: Milton Keynes, 1997, p. 108. With permission.)

carbon atom) in them. Over geological timescales (many millions of years) the main sources of atmospheric CO_2 are volcanic activity and oxidation of organic matter in buried sediments that have been brought to the Earth's surface. Volcanic activity includes outgassing of CO_2 from the mantle, and decarbonation, in which CO_2 is formed when carbonate and silicate rocks react when heated together at subduction zones (Equation 4.2). The geological timescale sinks for CO_2 are weathering of silicate rock (Chapter 2, Section 2.6), and conversion of CO_2 to organic matter by photosynthesis, followed by burial of the organic matter in sediments.

$$CaCO_{3(s)} + SiO_{2(s)} \longrightarrow CaSiO_{3(s)} + CO_{2(g)} \qquad (4.2)$$

On a short timescale (from days to a number of years) the principal carbon fluxes are CO_2 exchange between the atmosphere and land-based biomass, and CO_2 exchange between the atmosphere and surface waters of the oceans. CO_2 is taken up by photosynthesising biomass, and CO_2 is generated by the decomposition of organic carbon in **respiration** (Equation 4.3), which is the reverse process of photosynthesis.

$$nCH_2O_{(s)} + nO_{2(g)} \longrightarrow nCO_{2(g)} + nH_2O_{(g)} \qquad (4.3)$$

In the oceans, the uptake from and release of CO_2 to the atmosphere depends on the equilibria between atmospheric CO_2 and dissolved CO_2, which depends on the equilibria involving the various dissolved species of the carbonate system (Chapter 3, Section 3.3.3). This in turn is affected by photosynthesis and respiration of marine biomass, which takes up aqueous CO_2 from surface waters. The residence time for carbon in land-based plant biomass, the atmosphere, and the surface of the oceans is a few years, and is about 500 years in the deep oceans, which contrasts with a residence time of many millions of years for carbon in carbonate rock and deeply buried organic matter.

The principal anthropogenic sources of atmospheric CO_2 are the burning of fossil fuels and deforestation. This forms only about 3% of the short-term flux to the atmosphere, but it is not matched by a sink on an equivalent timescale, as these deposits of organic carbon were laid down over many millions of years. Some of this carbon is being taken up into the oceans, but the additional CO_2 flux from fossil fuels is leading to an increase in atmospheric CO_2 levels. This will be discussed further in Section 4.4.

4.3.2 THE NITROGEN CYCLE

Another important environmental cycle on earth is the **nitrogen cycle**, and a simplified version is shown in Figure 4.5. Elemental nitrogen, N_2, is a very **inert** (i.e., unreactive) diatomic gas that makes up 78% (by volume) of the atmosphere. The other principal gaseous forms of nitrogen are nitric oxide, NO; nitrogen dioxide, NO_2; and nitrous oxide, N_2O. Biological processes in the soil release all three gases, but especially N_2O, whereas lightning and biomass burning are major sources of NO and NO_2.

Although N_2 is exceedingly inert, a number of microorganisms, for example, *Rhizobium*, have the ability to break the N≡N bond to convert N_2 to ammonia, NH_3, in a process known as **fixation**. These microorganisms are found in blue-green algae

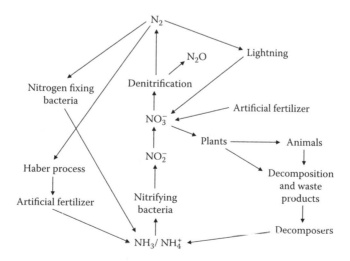

FIGURE 4.5 The nitrogen cycle. (From Dave McShaffrey. Ecosystems, www.Marietta.edu/~Biol/102/Ecosystems.html)

and lichens, and in the roots of legumes. The ammonia formed in fixation is used in biological processes during which it may be converted into other nitrogen-containing species such as nitrite, NO_2^-; nitrate, NO_3^-; amino acids; and proteins. The process of fixation is still not fully understood, but all microorganisms that can fix nitrogen contain an **enzyme** (a very large molecule that carries out chemical reactions in living organisms) called **nitrogenase**, which contains the element molybdenum, Mo. The overall scheme involves the oxidation of carbohydrate to carbon dioxide accompanied by the reduction of nitrogen to the ammonium ion (Equation 4.4; CH_2O is a general formula for plant organic matter). About 2×10^{11} kg of nitrogen is fixed per year by bacteria, whereas another 8.5×10^{10} kg of nitrogen is converted to ammonia in an industrial process called the Haber process, in which nitrogen is reacted directly with hydrogen at high pressure and temperature (Equation 4.5). Much of this ammonia is used for artificial fertilizer. A small amount of nitrogen is also fixed by lightning as nitric oxide, NO, and nitrogen dioxide, NO_2 (which is collectively known as NO_x).

$$3CH_2O_{(s)} + 3H_2O_{(l)} + 2N_{2(g)} + 4H^+_{(aq)} \longrightarrow 3CO_{2(g)} + 4NH^+_{4(aq)} \qquad (4.4)$$

$$3H_{2(g)} + N_{2(g)} \longrightarrow 2NH_{3(g)} \qquad (4.5)$$

Ammonia formed during fixation can be converted by plants to protein or it may be oxidised to nitrate as a souce of energy by **nitrifying bacteria** such as *Nitrosomonas* and *Nitrospira*. The formation of nitrate from ammonia is termed **nitrification**. This is a two-stage process, which is shown in Equations 4.6 and 4.7.

$$2NH^+_{4(aq)} + 3O_{2(g)} + 2H_2O_{(l)} \longrightarrow 2NO^-_{2(aq)} + 4H_3O^+_{(aq)} \qquad (4.6)$$

$$2NO^-_{2(aq)} + O_{2(g)} \longrightarrow 2NO^-_{3(aq)} \qquad (4.7)$$

Nitrate can be reduced back to nitrogen under both aerobic and anaerobic conditions by **denitrifying bacteria** such as *Paracoccus denitrificans* (Equation 4.8). These bacteria can also form nitrous oxide (N_2O), which is the major source of this gas in the atmosphere. N_2O is a cause for environmental concern because it is an extremely efficient greenhouse gas (Section 4.4) and it also contributes to ozone depletion (Section 4.5). The reduction processes involving nitrate are collectively called **denitrification**.

$$4NO^-_{3(aq)} + 5CH_2O_{(s)} + 4H_3O^+_{(aq)} \longrightarrow 2N_{2(g)} + 5CO_{2(g)} + 22H_2O_{(l)} \qquad (4.8)$$

During the decay of plants and animals, more complex organic nitrogen-containing molecules break down into simpler species, eventually forming ammonia in a process termed **ammonification**.

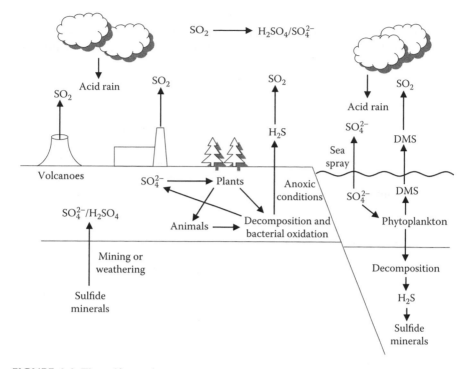

FIGURE 4.6 The sulfur cycle.

4.3.3 THE SULFUR CYCLE

Sulfur cycles through the environment, but its abundance in the atmosphere has greatly increased since the Industrial Revolution. This has been mainly due to the emission of sulfur dioxide, SO_2, as a consequence of the consumption of fossil fuels, although metal extraction and production of fertilizers has added to the cycling of sulfur in general. A simplified form of the sulfur cycle is shown in Figure 4.6. Most sulfur is found in the lithosphere and seawater, with minor amounts in the biosphere and trace amounts in the atmosphere. It has been established that in natural processes sulfur tends to be reduced, whereas in anthropogenic processes it tends to be oxidised.

Just as in the nitrogen cycle, microorganisms play an important role in the sulfur cycle, and the two cycles are similar in several aspects. For example both elements tend to be in their most reduced form when found in living organisms (i.e., as $-NH_2$ and $-SH$ groups). The oxidation of reduced forms of sulfur by oxygen can occur through the action of microorganisms in the soil, the sediment and the water column, whereas, within the atmosphere, chemical rather than biological oxidation occurs.

Sulfur is found in amino acids and proteins, and when these decompose the initial product is hydrogen sulfide, H_2S, but this is usually oxidised by sulfur-oxidising bacteria such as *Thiobacillus thiooxidans*. In anoxic conditions such as swamps and marine sediments, however, sulfur remains in its reduced form as hydrogen sulfide. When H_2S is released to the atmosphere, it is oxidised (Equation 4.9) whereas if it remains in the sediment, it may react with metal ions to form insoluble sulfides. Iron is relatively abundant in sediments and forms the major sulfide minerals troilite, FeS, and iron pyrites, FeS_2 (Chapter 3, Section 3.4.2). These minerals, together with decomposed organic matter, give a black colour to sediments.

$$H_2S_{(g)} \longrightarrow S_{(s)} \xrightarrow{O_2} SO_{2(g)} \xrightarrow{O_2} SO_{3(g)} \xrightarrow{H_2O} SO_{4(aq)}^{2-} \qquad (4.9)$$

Within the oceans the most important biological sulfur compound is dimethyl sulfonopropionate, which is produced by phytoplankton. This decomposes to yield dimethyl sulfide, DMS, which is released to the atmosphere, and provides a key link in the sulfur cycle. In the atmosphere, DMS is oxidised to sulfur dioxide (Equation 4.10), which ultimately forms sulfate. Sulfate particles are also directly released to the atmosphere from sea spray.

$$
\begin{array}{c}
\underset{\displaystyle H}{\overset{\displaystyle H}{H-C-S-C-H}} \xrightarrow{O_2} SO_2 + 2CO_2 + 3H_2O \\[2mm]
\qquad\qquad\qquad\quad \Big\downarrow O_2 \\[2mm]
\qquad\qquad\qquad SO_3 \xrightarrow{H_2O} H_2SO_4
\end{array}
\qquad (4.10)
$$

Although it is energetically unfavourable to reduce sulfate to sulfide, this does occur within the sulfur cycle. Microorganisms such as *Desulfovibrio desulfuricans* in marine and freshwater sediments use the energy released during the oxidation of carbohydrate to reduce sulfate, while sulfate acts as the oxygen source for the formation of carbon dioxide (Equation 4.11).

$$2CH_2O_{(s)} + SO_{4(aq)}^{2-} + 2H^+ \longrightarrow 2CO_{2(g)} + H_2S_{(g)} + 2H_2O_{(l)} \qquad (4.11)$$

Soil can act both as a sink and source of various inorganic and organic sulfur species. In humid regions, organic sulfur is concentrated in the topsoil, where it is oxidized to sulfate, which can then be either leached into the subsoil or taken up by vegetation. Inorganic sulfur is invariably in the form of sulfate bound in the subsoil, unless the region is arid, in which case gypsum, $CaSO_4.2H_2O$ occurs as a coating on the topsoil. On agricultural land the demand for sulfur is high and many fertilizers incorporate ammonium sulfate, $(NH_4)_2SO_4$, as the sulfur source.

4.4 GLOBAL WARMING AND THE GREENHOUSE EFFECT

Based on the amount of solar radiation falling on the Earth's surface, the mean surface temperature of the Earth should be 255 K, well below the freezing point of water. The atmosphere retains incident solar energy and consequently keeps the

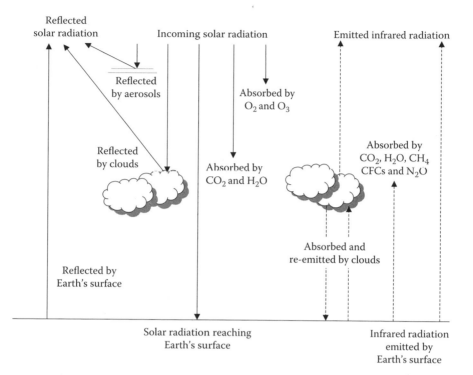

FIGURE 4.7 The greenhouse effect.

planet warm, in what is called the **greenhouse effect** (Figure 4.7). Ultraviolet, visible and infrared radiation from the Sun reaches the Earth's atmosphere, where short-wavelength UV radiation is absorbed by O_2 and O_3 in the upper atmosphere. Some longer-range radiation is absorbed by CO_2 and H_2O, whereas some is reflected off the tops of clouds or from the Earth's surface. The fraction of reflected light is called the **albedo** and is over 0.5 for clouds but less than 0.1 for oceans. The average for the planet is 0.3, which means that about 30% of incoming solar radiation is reflected back into space. The solar radiation that is not absorbed by the atmosphere or reflected directly back is absorbed by the Earth's surface, warming it. As a result of this warming, the Earth's surface emits long-wavelength infrared radiation back into the atmosphere. In the atmosphere there are a number of **greenhouse gases** that are transparent to ultraviolet and visible wavelength radiation but strongly absorb radiation at infrared wavelengths. These gases include carbon dioxide, CO_2, methane, CH_4, nitrous oxide, N_2O, and chlorofluorocarbons, CFCs, which are present in trace levels (ppm to ppb), as well as water, H_2O, which is present at levels of about 0.5–4%. The absorption of infrared radiation by these gases warms the Earth's atmosphere. Clouds also absorb and re-emit infrared radiation, some of it back to Earth's surface. It is the absorption of the infrared radiation emitted by the Earth's surface that causes the additional warming of the atmosphere, and hence of the Earth's surface that is the greenhouse effect. Figure 4.8 shows the various absorption bands of CO_2, H_2O, and CH_4 in the mid-infrared region (2.5–25 μm), which covers

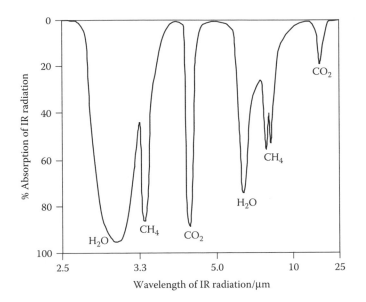

FIGURE 4.8 Absorption of infrared radiation by greenhouse gases.

much of the range of wavelengths emitted by the Earth. The intensity of the re-emitted infrared radiation peaks around 14 μm.

For several years reports of **global warming** have been appearing in both the popular press and scientific journals. Global warming is caused by an **enhanced greenhouse effect,** in which increasing concentrations of greenhouse gases in the atmosphere as a result of human activity, in particular CO_2, produced in the burning of fossil fuels, are leading to an increase in the Earth's temperature above the level it would be if no human activity were taking place. Figure 4.9 shows the increase in atmospheric carbon dioxide levels since 1959, and Table 4.2 shows the increase since the Industrial Revolution in atmospheric levels of four greenhouses gases. N_2O is produced in denitrification reactions mediated by bacteria (Section 4.3.2). It is emitted from soil, and atmospheric N_2O levels are increasing, probably because of widespread use of nitrate fertilisers. Methane is formed in rice paddies and is emitted by cattle. Both of these compounds are more powerful greenhouse gases than CO_2, as indicated by their **greenhouse warming potential**, which is the strength of a greenhouse gas relative to CO_2 (Table 4.2). As you can see, methane is 23 times more powerful as a greenhouse gas than carbon dioxide, and nitrous oxide is 296 times more powerful than carbon dioxide. Current estimates suggest that an average temperature rise of about 4°C will occur over the next 100 years.

Figure 4.10 shows global levels of carbon dioxide emissions, broken down into industrial activity and deforestation. The total global carbon dioxide emission are in excess of 22×10^{12} kg per year (22 billion tonnes per year). Industrial activity accounts for about 18×10^{12} kg per year (18 billion tonnes per year), and currently North America, Europe, and Asia generate the highest industrial emissions (about 83% of the total). Deforestation adds another 4 billion tonnes per annum, the main regions responsible for this being South America, Asia, and Africa. The combination

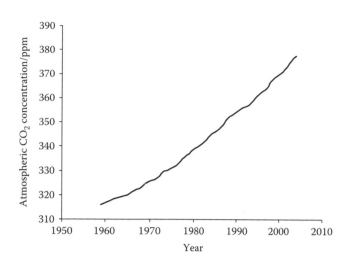

FIGURE 4.9 Increase in atmospheric carbon dioxide since 1959.

TABLE 4.2
Increase in Atmospheric Levels of Greenhouse Gases Since the Industrial Revolution

Gas	Greenhouse Warming Potential	Atmospheric Level			
		1700	1800	1900	2000
Carbon dioxide/ppm	1	280	280	300	370
Methane/ppb	23	750	770	1000	1750
Nitrous oxide/ppb	296	270	275	285	310
CFC/ppb	4600 (CFC-11)	0	0	0	0.677

of burning fossil fuels and deforestation should lead to a 0.75% increase in CO_2 levels per year based on the amount of carbon burned or not sequestered in trees. At the current level of atmospheric CO_2 of about 382 ppmv, this should lead to an increase of about 2.8 ppmv per year. The actual increase is lower than this, at about 1–2 ppmv per year, because the oceans are acting as a CO_2 sink. This itself, however, will cause problems because acidification of the oceans will particularly affect organisms that form $CaCO_3$ skeletons because the solubility of $CaCO_3$ increases with decreasing pH (Chapter 3, Section 3.3.3). In the following Self-Assessment Question (SAQ) you will work through how much CO_2 is emitted by one fuel tank's worth of petrol (gasoline) in a typical small family car.

SELF-ASSESSMENT QUESTION

Q4.2 The fuel tank of a typical family car can hold 42 L of petrol, which is equivalent to 29.5 kg of petrol. (The density of octane is 0.703 g mL^{-1}.)

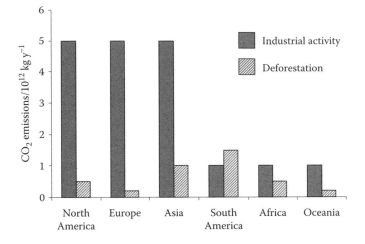

FIGURE 4.10 Global carbon dioxide emissions.

(i)   Assuming that petrol (gasoline) has a molecular formula of C_8H_{18}, use the periodic table to calculate the molar mass of petrol, and hence calculate how many moles there are in 29.5 kg of petrol.

(ii)   For every one mole of petrol that is burned, 8 moles of CO_2 are formed according to the equation $C_8H_{18} \rightarrow 8CO_2$. How many moles of CO_2 are formed by burning 29.5 kg of petrol?

(iii) Calculate the molar mass of CO_2, and hence calculate the mass of CO_2 formed. Express this in units of kilograms.

In 1992 the United Nations Framework on Climate Change (UNFCCC) was formed at the Rio Earth Summit to negotiate a worldwide agreement on reducing greenhouse gases. In 1997 the Kyoto Protocol was agreed upon, in which developed nations were required to cut their greenhouse gas emissions by 5.2% of their 1990 levels by 2008–12. In 2001 at Bonn, 186 countries (but not the U.S.) ratified and signed the Kyoto Protocol. It became an international law in 2005 when Russia signed, although the level of emission cuts were dropped from 5.2% to 1-3%. The EU has set itself a tight limit of an 8% cut on its 1990 levels. The U.K. legal target is 12.5%, to allow poorer EU countries room for development. Whether any of these targets will be achieved remains to be seen.

    Another approach being considered to reduce levels of carbon dioxide in the atmosphere is to sequester carbon dioxide in underground reservoirs such as depleted oil and gas reservoirs, and in saline formations (reservoirs of salt-laden water). At the high pressures that are found at depths below about 800–1000 m, CO_2 is significantly compressed and occupies much less volume than as a gas at atmospheric pressure. It has been suggested that CO_2 could be stored as a compressed gas in porous rocks under the same cap rocks that trap oil and gas, or dissolved in aqueous solution in similar reservoirs, or in solid form by reaction with minerals to form metal carbonates or related minerals.

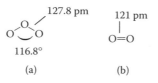

FIGURE 4.11 The structure of ozone, showing (a) the ozone molecule, O_3; and (b) the dioxygen molecule, O_2.

4.5 THE OZONE LAYER

Ozone is an **allotrope** (i.e., a chemical form) of oxygen that has three oxygen atoms in each molecule, giving it a molecular formula of O_3. Figure 4.11 shows the chemical structure of ozone and dioxygen, O_2, and you can see that O_3 is a bent molecule with longer bonds than O_2. (We will use the name dioxygen in this section to indicate when we specifically mean O_2 as opposed to O_3.) The greater length of the O–O bond in O_3 (127.8 pm) compared to O_2 (121 pm) indicates that the bond is a little weaker, which is key to its beneficial (and harmful) properties. The presence of ozone in the atmosphere is, as we will see shortly, essential to life on land.

The vertical distribution of O_2 and O_3 in the atmosphere is not uniform. At the outer reaches of the atmosphere, oxygen is in the form of atomic oxygen (O) because intense radiation from the sun breaks down O_2 molecules. About 60 km from the Earth's surface, there are increasing concentrations of dioxygen (O_2) and nitrogen (N_2). Ozone is concentrated in a layer between 20 and 30 km from the Earth's surface in the stratosphere, as shown in Figure 4.1, its maximum concentration of about 1% occurring 25 km from the Earth's surface. Ozone also occurs at ground level because of air pollution, which you will learn about in Section 4.6.

4.5.1 OZONE FORMATION AND UV PROTECTION

Ozone is formed by the action of ultraviolet (UV) light from the Sun on O_2, which **photolyses** the molecule, splitting it into oxygen atoms (Equation 4.12). These are very reactive and can combine with another O_2 molecule (Equation 4.13). In this reaction, a third body, M (usually a nitrogen molecule), is required to remove excess energy.

$$O_{2(g)} + \text{UV radiation} \longrightarrow O_{(g)} + O_{(g)} \qquad (4.12)$$

$$O_{(g)} + O_{2(g)} + M \longrightarrow O_{3(g)} + M \qquad (4.13)$$

The presence of the ozone layer is essential to land-based life because it provides a barrier to ultraviolet radiation. It provides this by absorbing UV radiation, causing the molecule to break up (Equation 4.14). The free oxygen atom thus formed can react with another module of ozone (Equation 4.15).

$$O_{3(g)} + UV \text{ radiation} \longrightarrow O_{(g)} + O_{2(g)} \qquad (4.14)$$

$$O_{(g)} + O_{3(g)} \longrightarrow 2O_{2(g)} \qquad (4.15)$$

Exposure to UV radiation in the 200–300 nm range is known to promote skin cancer, which is due to DNA damage. This can be better understood if we do a quick calculation. The energy of radiation, E, in kJ mol^{-1}, which has wavelength λ in metres, is given by Equation 4.16. h is Plank's constant (6.626×10^{-34} J s), c is the speed of light (2.998×10^8 m s^{-1}) and N_A is the Avogadro constant.

$$E = \frac{h \times c \times N_A}{1000 \times \lambda} \qquad (4.16)$$

If we put in a value of 290×10^{-9} m for λ (i.e., a wavelength of 290 nm), we obtain a value of E of 412 kJ mol^{-1}, which is enough energy to break many chemical bonds such as C-O and C-C (see Table 1.8. for a list of bond energies). Ozone absorbs UV radiation over the crucial range of about 280–310 nm, but its maximum absorption occurs at about 255 nm. If it did not absorb radiation over this wavelength range, much more damage to DNA and other biological molecules would occur. Shorter (higher-energy) wavelengths of radiation are cut out by O_2, because the O=O bond in O_2 is shorter than in O_3, and therefore it has a higher bond energy, and is able to absorb high-energy radiation. UV radiation with a wavelength longer than 310 nm does less damage than shorter-wavelength radiation as it does not have enough energy to break as many chemical bonds.

SELF-ASSESSMENT QUESTIONS

Q4.3   (i)   The maximum absorption of UV radiation by ozone occurs at 255 nm. Express this wavelength in units of metres, m, using scientific notation.

(ii)   Calculate the energy of radiation absorbed by ozone at this wavelength.

(iii) Name one chemical bond from those listed in Table 1.8 that would be broken by radiation of this wavelength, if the radiation were not absorbed by ozone.

Q4.4   (i)   The bond energy of dioxygen, O_2, is 498 kJ mol^{-1}. If O_2 absorbs UV radiation of this energy, what is the wavelength of radiation being absorbed in units of metres, m?

(ii) Express this wavelength in units of nanometres, nm.

4.5.2 OZONE DEPLETION

Ozone is readily detected owing to its strong absorption of UV radiation between 200 and 300 nm. However, accurate determination is difficult because of seasonal variations and solar emission cycles. Monitoring of ozone began in the early 1970s, and concern over the ozone layer was first expressed in the mid-1970s. In 1985 atmospheric analysis showed a very serious depletion of ozone over Antarctica, which gave rise to the concept of an ozone hole. The cause was found to be chlorofluorocarbons (CFCs). These were introduced in the 1930s as replacement refrigerants for liquid ammonia or sulfur dioxide, and they had also been used for a number of years as propellants and in the formation of foamed plastics. Studies at the time showed them to be stable to all reactions carried out in normal atmospheres. As they gained widespread use, CFCs began to concentrate in the troposphere, where their chemical stability meant that they were not easily removed. Over time, CFCs diffused into the stratosphere, where the ozone layer is situated.

It is now known that CFCs **photolyse** (are broken down by ultraviolet radiation) in the stratosphere to yield chlorine atoms, $Cl\cdot$. A chlorine atom is an example of a **radical**, a chemical species with an unpaired electron, which is indicated by the $\cdot$ symbol. The presence of unpaired electrons means that radicals are usually very reactive. A chlorine radical can attack an ozone molecule to yield a chlorine monoxide radical ($ClO\cdot$) and dioxygen (O_2, Equation 4.17), and the $ClO\cdot$ thus produced can react with an O atom to form O_2 and $Cl\cdot$ (Equation 4.18). The net effect is that an ozone molecule is destroyed (Equation 4.19). The $Cl\cdot$, which is regenerated, can then repeat the cycle, destroying more ozone.

$$Cl\cdot_{(g)} + O_{3(g)} \longrightarrow ClO\cdot_{(g)} + O_{2(g)} \tag{4.17}$$

$$ClO\cdot_{(g)} + O_{(g)} \longrightarrow Cl\cdot_{(g)} + O_{2(g)} \tag{4.18}$$

$$O_{3(g)} + O_{(g)} \longrightarrow 2O_{2(g)} \tag{4.19}$$

Not all CFCs can attack ozone to the same degree, and a concept called the **ozone depletion potential** has been introduced. The most aggressive CFC (CFC-11) is given an ODP value of 1.0, and all other CFCs have values below this (for example, CFC-12 has an ODP of 0.9). CFC replacements are being developed that have ODPs of around 0.013. It is now known that bromine also takes part in ozone depletion reactions in the stratosphere, accounting for 20% of the total depletion rate.

Global production of the two most common CFCs (CFC-11 and CFC-12) rose rapidly from 50×10^6 kg y^{-1} (50,000 tonnes per annum) in 1950 to 725×10^6 kg y^{-1} (725,000 tonnes per annum) in 1976. About 90% of this was released directly into the atmosphere. Although levels of CFCs are very small (1 ppb), they have a long lifetime, and one CFC molecule can destroy thousands of ozone molecules. The use of CFCs in the EU and the U.S. is now banned.

SELF-ASSESSMENT QUESTION

Q4.5   The molar mass of CFC-12 is 120.91 g mol^{-1}. If one molecule of CFC-12 destroys on average 1000 molecules of ozone, what mass of ozone will be destroyed by 1 kg of CFC-12?

TABLE 4.3
Lifetimes of CFCs

CFC	Formula	Lifetime/Year
CFC-11	$CFCl_3$	77
CFC-12	CF_2Cl_2	139
CFC-13	$C_2F_3Cl_3$	92
Carbon tetrachloride	CCl_4	76
Methyl chloroform	CH_3CCl_3	8.3
HCFC-22	CHF_2Cl	22
HFC-134a	$C_2H_2F_4$	10
Halon-1301	CF_3Br	101
Halon-1211	CF_2BrCl	12.5

Source: From "Stratospheric Ozone," UK Strato-spheric Ozone Review Group, Her Majesty's Sta-tionary Office, London, 1988. With permission.

The UN Convention on the Protection of the Ozone Layer was agreed upon in 1985 in Vienna, modified in 1987 (Montreal) and 1990 (London). More than 70 countries are signatories, and CFCs, CCl_4, and methyl chloroform (CH_3CCl_3) will be phased out by 2005. However, repair to the atmosphere will be slow owing to the lifetimes of the CFC compounds, which are of the order of decades rather than years (Table 4.3).

4.6   AIR POLLUTION

Air pollution has been a problem for as long as humans have been cooking over open fires (and domestic-scale pollution from this source is still a problem in some regions of the world), but the scale of air pollution has increased since the start of the Industrial Revolution. Emissions to the atmosphere that have a global impact include carbon dioxide (Section 4.4) and CFCs (Section 4.5), whereas emissions that have a regional, local, or domestic effect include sulfur dioxide (SO_2), nitrogen oxides (NO_x), carbon monoxide, volatile and semivolatile organic compounds (VOCs and SVOCs), particulates, and radon gas. **Primary pollutants** are emitted directly into the atmosphere, whereas **secondary pollutants** are formed from primary pollutants in atmospheric reactions. Some pollutants are largely or entirely anthropogenic in origin, for example, ground-level ozone, which is produced as a result of vehicle emissions; some have both anthropogenic and natural sources, such as sulfur dioxide, which comes both from volcanoes and the burning of fossil fuels; whereas a few pollutants are largely or wholly natural in origin, for example, radon gas produced by radioactive decay (although radon is mainly a problem because of anthropogenic activities, i.e., living and working in buildings).

4.6.1   THE KEY POLLUTANTS

In this section we will consider the important anthropogenic atmospheric pollutants, excluding CO_2 and CFCs, which have already been considered, and we will start with sulfur dioxide, SO_2. Sulfur dioxide is a major contributor to acid rain (Box

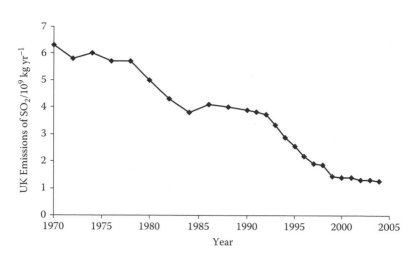

FIGURE 4.12 Emissions of sulfur dioxide by the U.K.

3.1), which is particularly an issue in regions and countries that lie around and downwind of (based on prevailing winds) major sources of the gas. SO_2 can be oxidised in the atmosphere to form sulfur trioxide, SO_3, which combines with water to form droplets of sulfuric acid, H_2SO_4 (Equation 4.20). Sulfuric acid is one of the two main contributors to acid rain, rain that has a pH value below its natural level of pH 5.6 (Chapter 3, Section 3.3.3).

$$H_2S_{(g)} \longrightarrow S_{(s)} \xrightarrow{O_2} SO_{2(g)} \xrightarrow{O_2} SO_{3(g)} \xrightarrow{H_2O} H_2SO_{4(aq)} \quad (4.20)$$

Anthropogenic sources of sulfur dioxide derive mostly from the burning of fossil fuels and from smelting of metal sulfide ores such as chalcopyrite, $CuFeS_2$, which is used in copper production. In coal the sulfur takes the form of pyrites (FeS_2) and organic sulfur compounds. Coal contains about 1.5% sulfur on average, leading to SO_2 emission levels of around 1200 ppm, but values may be as high as 7% for some types of coal. Sulfur can be removed from gas and every grade of oil except very heavy fuel oils. Typically, these heavy fuel oils may contain up to 3% sulfur. Controls on emissions from factory and power station chimneys, and the introduction of low-sulfur fuels are leading to reductions in sulfur emissions in developed countries (for example, in the U.K. as shown in Figure 4.12), but this is being counterbalanced by rapidly increasing use of fossil fuels in developing countries that lack controls on emissions and that have not introduced low-sulfur fuels.

Another aspect of pollution caused by SO_2 emissions is an increase in atmospheric **aerosols** (solid particles or liquid droplets suspended in the atmosphere) by the formation of droplets of H_2SO_4 or fine particles of sulfate, SO_4^{2-}. Aerosols in the stratosphere cause cooling of this part of the atmosphere by reflecting solar radiation back into space, reducing the amount that is absorbed by stratospheric O_2 and O_3. The overall effect that this will have on global climate is unclear because of the complex interplay between the stratosphere and troposphere, but short-term (a year or two) global cooling has occurred following volcanic eruptions such as Mount Pinatubo that put sulfate aerosols into the stratosphere (1991).

Nitric oxide and nitrogen dioxide (NO_x) are major regional and local pollutants. Anthropogenic sources of NO and NO_2 include the burning of biomass, but a major source of NO_x is the combustion of fossil fuels in power stations and internal combustion engines. There are two main routes to the formation of NO_x from fossil fuels. One route involves combining atmospheric nitrogen and oxygen at high temperatures (thermal NO_x, Equation 4.21). The other route involves the combustion of nitrogen compounds originally present in the fuel, and factors such as burner design will determine which is the dominant source. Typical flue gas concentrations of NO_x for a coal-fired power station are about 550 ppm. NO can be oxidised in the atmosphere to NO_2 by oxygen (Eqn 4.22), or other atmospheric species such as ozone, O_3 (Eqn 4.23), or oxygen-containing radicals such as **peroxides**, ROO·. A very serious local effect of the formation of NO in vehicle engines is photochemical smog, which will be considered in the next section.

$$N_{2(g)} + O_{2(g)} \longrightarrow 2NO_{(g)} \tag{4.21}$$

$$2NO_{(g)} + O_{2(g)} \longrightarrow 2NO_{2(g)} \tag{4.22}$$

$$NO_{(g)} + O_{3(g)} \longrightarrow NO_{2(g)} + O_{2(g)} \tag{4.23}$$

NO_2 formed in the atmosphere can dissolve in water to give droplets of nitric acid, HNO_3, which, together with H_2SO_4, are the two main contributors to acid rain. The level of NO_2 emissions in the U.K. since 1970 can be seen in Figure 4.13, and this shows that since 1990, NO_2 levels have dropped significantly following the introduction of catalytic converters in cars, which convert to harmful to NO_2 harmless N_2.

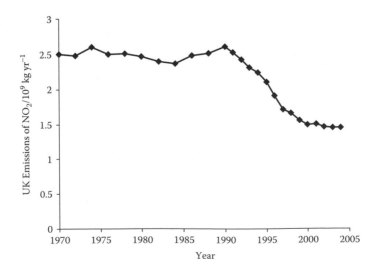

FIGURE 4.13 Emissions of nitrogen dioxide by the U.K.

Two other nitrogen-containing gaseous pollutants are nitrous oxide, N_2O, and ammonia, NH_3. Ammonia is released from waste treatment plants, biological decay, and animal excrement, and can combine with sulfuric acid to form ammonium hydrogen sulfate aerosols (Equation 4.24). It is also a minor contributor to acid rain as a precursor molecule. N_2O is a powerful greenhouse gas. The main source of N_2O is due to denitrifying bacteria in the soil (Section 4.3.2), and N_2O levels in the atmosphere have increased with the increase in nitrate fertilizer being added to soils.

$$NH_3 + H_2SO_4 \longrightarrow NH_4SO_4 \qquad (4.24)$$

Methane, CH_4, is another powerful greenhouse gas. Anthropogenic sources include anaerobic fermentation of organic material in rice paddies, as well as enteric fermentation in the digestive systems of ruminants (e.g., cows). Methane is also released in coal burning, gas extraction, and biomass burning.

Carbon monoxide, CO, is poisonous at relatively low levels, and is formed by burning carbon-based fuels in a limited air supply. This causes problems at the domestic level by carbon monoxide poisoning (Section 4.6.3) and at local level as part of the process of smog formation (Section 4.6.2).

Volatile organic compounds (VOCs) are organic compounds that have appreciable or even quite low vapour pressures but that are sufficient to produce measurable quantities in the atmosphere. Some VOCs, mainly fairly low-molecular-weight compounds used as solvents similar, are particularly a concern with regard to indoor air quality in factories, offices, homes, etc. (see Section 4.6.4). Higher-molecular-weight compounds that cause concern include polychlorinated biphenyls (PCBs). PCBs are a global problem because of the stability and persistence of these molecules. Air circulation patterns have carried PCBs produced in mid latitudes in industrial countries to high polar latitudes where they have been found to have accumulated in the bodies of animals such as seals and polar bears.

Particulates are serious pollutants on domestic, local and regional scales. One of the main anthropogenic sources is combustion. Coal generally contains a proportion of silicate-based material, which forms fine particles of **fly ash** when the coal is burned. Levels of ash in U.K. solid fuels range from 5 to 20% by weight. Emissions of pulverized fly ash from existing power stations is limited in the U.K. to 115 mg m^{-3}, which requires the removal of over 99% of the ash originally present in the flue gases. The limit for new stations is 50 mg m^{-3}.

The incomplete combustion of carbon-based fuels leads to the formation of soot, which essentially consists of fine particles of unburned carbon. This is often accompanied by the formation of CO, and occurs when carbon-based fuel is burned in a limited air supply. The diameter of these particles can range from 100 μm down to 10 nm, with the major particles in the 0.1–10 μm range. The smallest nuclei are formed by condensation of hot vapours that undergo rapid coagulation to form larger particles. Coarse particles are normally 2 μm or larger. Particulates are defined by their size, so that PM_{10} are particles below 10 μm, whereas $PM_{2.5}$ are particles below 2.5 μm. Figure 4.14 shows estimated U.K. emissions of primary PM_{10} particulate matter as a percentage by source category in 1998. Automotive particulate materials

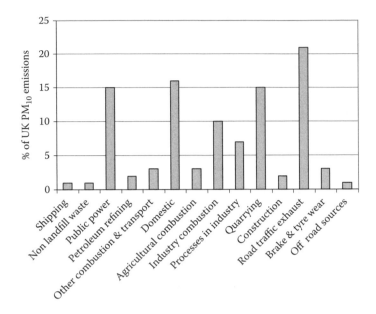

FIGURE 4.14 PM$_{10}$ Emssion Data for the U.K. for 1998.

can contain a variety of organic substances, including polyaromatic hydrocarbons (PAHs), of which benzo(a)pyrene is an example, PCBs, polychlorinated dibenzo-dioxins, and dibenzo-furans (PCDD and PCDF, Chapter 2, Section 2.7.3). We will discuss the effects of unburned carbon in the formation of smog in the next section.

Hydrogen chloride, HCl, can be produced by the combustion of coal, which can contain up to 0.2% chlorides. HCl can also be produced during the incineration of the plastic polyvinylchloride, PVC. Although not considered as important as other emissions, HCl can add to the acidity of precipitation.

Probably the most prevalent form of air pollution not necessarily connected with combustion is dust. Although much soil-derived dust is natural, agricultural activity has lead to considerable soil erosion and a significant proportion of this material gets into the atmosphere. Particulate matter in addition to soot can be emitted in many industrial processes such as metal smelting, construction, quarrying, etc. It is convenient to distinguish between different sizes of particles, and in the U.K., particles are classed as **grit, dust,** or **fume**. Grit consists of large particles (>76 μm diameter), which will rapidly settle out of the air because of gravity and are just visible to the naked eye; dust consists of intermediate-sized particles (76 μm > diameter >1 μm), which can be seen in an optical microscope; and fume consists of very small particles (<1 μm diameter), which can remain suspended in air for long periods and which are just visible by electron microscopy. In the U.K., emissions of grit, dust, and fume from various processes are controlled under the Clean Air Act and the Environmental Protection Act 1990. Suspended particulate matter is very diverse in character, and includes both organic and inorganic substances.

A breakdown of the total annual U.K. emissions of black smoke (soot), SO$_2$, NO$_x$, CO, and VOCs is given in Table 4.4.

TABLE 4.4
Annual U.K. Emissions of Air Pollutants (Values given in 10^6 kg)

Source	Black Smoke	SO_2	NO_x	CO	VOC (excluding CH_4)
Domestic	191	135	68	339	50
Commercial/industrial	92	683	337	342	43
Power stations	25	2644	785	47	12
Refineries	2	109	36	1	0
Processes and solvents	0	0	0	0	1059
Road vehicles (petrol)	15	22	702	5649	590
Road vehicles (diesel)	182	30	596	102	172
Railways	0	3	32	12	8
Forests	0	0	0	0	80
Gas leakage	0	0	0	0	34
Other	5	65	144	31	18
Total	512	3691	2700	6523	2066

Source: From "Digest of Environmental Protection and Water Statistics," No. 13, Her Majesty's Stationary Office, London, 1990. With permission.

4.6.2 URBAN SMOG

The word **smog** is formed from a combination of smoke and fog, and was coined to describe a type of air pollution occurring in urban environments. The two basic types of smog are **classical smog** and **photochemical smog**; we will consider the causes and effects of classical smog first. During an ideal combustion process, coal or other carbon-based fuel (represented by a generic molecular formula of CH) burns yielding only carbon dioxide and water (Equation 4.25). However, an ideal situation is rarely found and carbon monoxide and carbon soot (i.e., carbon particulates) can be formed if the oxygen supply is limited (Equations 4.26 and 4.27). To make matters worse, coal and fuel oil contain sulfur as an impurity in levels up to a few percent, much of it in the form of iron pyrites, FeS_2. When this is burnt, it yields sulfur dioxide (Equation 4.28). Thus, burning of coal or fuel oil will yield carbon dioxide, carbon monoxide, sulfur dioxide, and particulate carbon (smoke), the latter three having a direct impact on people's health.

$$4CH_{(s)} + 5O_{2(g)} \longrightarrow 4CO_{2(g)} + 2H_2O_{(l)} \qquad (4.25)$$

$$4CH_{(s)} + 3O_{2(g)} \longrightarrow 4CO_{(g)} + 2H_2O_{(l)} \qquad (4.26)$$

$$4CH_{(s)} + O_{2(g)} \longrightarrow 4C_{(s)} + 2H_2O_{(l)} \qquad (4.27)$$

$$4FeS_{2(s)} + 11O_{2(g)} \longrightarrow 8SO_{2(g)} + 2Fe_2O_{3(s)} \qquad (4.28)$$

The combination of smoke and associated gases formed from the burning of coal and fuel oils leads to classical smog. In the U.K., the densest fogs form in calm, damp, cold conditions in early mornings in winter; this, combined with the need for heating at this time of year, made the winter months the most liable for classical smog. The most notorious smog of this type occurred over 5–8 December 1952, in London, as a result of which 4000 people died. There was no wind and temperatures were around freezing with a relative humidity of over 85%. When sulfur dioxide and smoke levels exceeded 500 g m^{-3}, people began to be affected by bronchial irritation and asthma, and when they reached 750 g m^{-3}, increases in sudden deaths were recorded in hospitals. Final peak levels of sulfur dioxide and smoke reached a staggering 4000 g m^{-3}. Essentially, what formed in London during those 3 days was an aerosol of sulfuric acid (formed via the oxidation of sulfur dioxide, Equation 4.20) that started to dissolve lung linings, and many people suffered respiratory disease long after the smog lifted. This type of smog incident lead to the Clean Air Act in the U.K. 1956, and the introduction of "smokeless fuels."

The second type of smog is photochemical smog. This is formed in a very different way from classical smog, and occurs under very different climatic conditions. In particular, it forms in strong sunlight in the presence of emissions from petrol engines, and so tends to occur most widely in large cities at low latitudes in summer daytime. The presence of a large number of motor vehicles results in the emission of large amounts of exhaust gases, which contain volatile hydrocarbons (unburned petrol or gasoline) and NO_x. NO_2 is photolysed by UV radiation, forming NO and an oxygen atom, and this oxygen atom can combine with an oxygen molecule to form ozone (Equation 4.29). When ozone is formed in the ozone layer high in the Earth's atmosphere, it provides protection from UV radiation, but at ground level, it causes major health and environmental problems. Ozone can be photolysed to form O_2 and a high-energy oxygen atom, and reaction of this high-energy oxygen atom with a water molecule forms hydroxyl radicals, OH· (Equation 4.30). Overall, one NO_2 molecule forms two OH· radicals.

$$NO_2 \xrightarrow{\text{UV radiation}} NO+O \xrightarrow{O_2} NO+O_3 \qquad (4.29)$$

$$O_3 \xrightarrow{\text{UV radiation}} O_2+O \xrightarrow{H_2O} 2HO\cdot+O_2 \qquad (4.30)$$

The hydroxyl radicals can react with volatile hydrocarbons from car exhausts to form hydrocarbon radicals (Equation 4.31), and these in turn can form peroxy radicals (Equation 4.32), alkoxy radicals (Equation 4.33), and aldehydes (Equation 4.34). One particular family of compounds that causes problems are PANs, which are formed from aldehydes via acylperoxy radicals (Equations 4.35 and 4.36). The first member of the family is peroxyacetyl nitrate (hence, PAN), in which the R group shown in Equation 4.33 is a methyl group, CH_3. PANs are eye irritants and are also phytotoxic (i.e., they are toxic to plants and can therefore cause crop damage). If there is an air inversion (i.e., cold air sits below warm air, and there is little mixing), then the concentrations of PAN can exceed 50 ppb. Effects of the photochemical smog were first reported in the Los Angeles basin in 1944, and the problems with vegetation have continued since then, especially in the Californian forests.

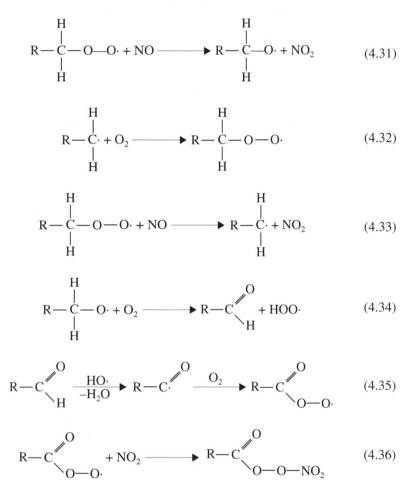

$$R—\overset{\overset{\displaystyle H}{|}}{\underset{\underset{\displaystyle H}{|}}{C}}—O—O\cdot + NO \longrightarrow R—\overset{\overset{\displaystyle H}{|}}{\underset{\underset{\displaystyle H}{|}}{C}}—O\cdot + NO_2 \qquad (4.31)$$

$$R—\overset{\overset{\displaystyle H}{|}}{\underset{\underset{\displaystyle H}{|}}{C}}\cdot + O_2 \longrightarrow R—\overset{\overset{\displaystyle H}{|}}{\underset{\underset{\displaystyle H}{|}}{C}}—O—O\cdot \qquad (4.32)$$

$$R—\overset{\overset{\displaystyle H}{|}}{\underset{\underset{\displaystyle H}{|}}{C}}—O—O\cdot + NO \longrightarrow R—\overset{\overset{\displaystyle H}{|}}{\underset{\underset{\displaystyle H}{|}}{C}}\cdot + NO_2 \qquad (4.33)$$

$$R—\overset{\overset{\displaystyle H}{|}}{\underset{\underset{\displaystyle H}{|}}{C}}—O\cdot + O_2 \longrightarrow R—C\overset{\displaystyle O}{\underset{\displaystyle H}{\diagdown}} + HOO\cdot \qquad (4.34)$$

$$R—C\overset{\displaystyle O}{\underset{\displaystyle H}{\diagdown}} \xrightarrow[-H_2O]{HO\cdot} R—C\overset{\displaystyle O}{\cdot} \xrightarrow{O_2} R—C\overset{\displaystyle O}{\underset{\displaystyle O—O\cdot}{\diagdown}} \qquad (4.35)$$

$$R—C\overset{\displaystyle O}{\underset{\displaystyle O—O\cdot}{\diagdown}} + NO_2 \longrightarrow R—C\overset{\displaystyle O}{\underset{\displaystyle O—O—NO_2}{\diagdown}} \qquad (4.36)$$

SELF-ASSESSMENT QUESTION

Q4.6   Shanghai lies at a latitude of about 31° N. Will the city experience worse photochemical smogs in December or June each year?

4.6.3 DISPERSAL OF POLLUTANTS

The airborne dispersion of pollutants is dependant on several factors but, in general, low wind speeds result in high pollutant concentrations and vice versa. Air pollution from small-scale emissions may be localised to the immediate vicinity of the source, but large-scale emissions such as those from power stations may cause regional or global problems.

TABLE 4.5
Time and Distance Dispersal of Pollutants

Time of Travel	Distance	Affected Area
Hours	10's km	Boundary layer
Days	1000's km	Entering troposphere
Weeks	Global	Troposphere in one hemisphere
Months	Global	Troposphere in both hemispheres

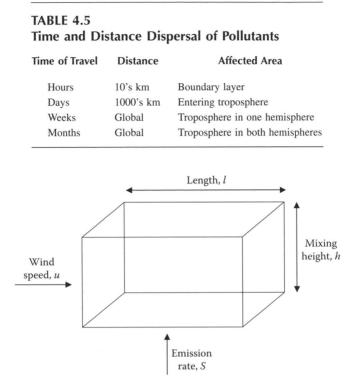

FIGURE 4.15 The box method for estimating airborne pollution. (From Roy Harrison, *Understanding Our Environment,* Royal Society of Chemistry, 1992, p. 30. With permission.)

Vertical mixing of pollutants in the boundary layer (Section 4.1) is largely determined by the atmospheric stability related to buoyancy effects. As a generalisation, mixing within the boundary layer is rapid; thus, pollutants can be spread quickly within its 1-km thickness. Table 4.5 shows the time and distance scales involved in the dispersion of pollutants emitted from the ground. A simple **box** approach (Figure 4.15) can be used to estimate the concentration of pollutants at a particular location at a given time, in which the concentration, C, of a pollutant depends on its emission rate, S; the length of the Box, l; the wind speed, u; and the mixing height, h of the atmospheric gases (Equation 4.37). The concentration of pollutants in an urban area, however, does not necessarily increase with decreasing wind speed as rapidly as predicted. Wind speeds tend to drop at night and rise again in the morning (especially near the coast where afternoon sea breezes occur), but so do emissions (fewer factories working, fewer cars on the road etc.).

$$C = \frac{S \times l}{u \times h} \qquad (4.37)$$

SELF-ASSESSMENT QUESTION

Q4.7 A factory is emitting SO_2. Using the box approach, calculate the concentration of SO_2 being emitted.

  (i)   The wind speed is 15 km hr^{-1} and the emission rate of SO_2 is 5 kg hr^{-1}. Convert these to SI units (i.e. m s^{-1} and g s^{-1}).

  (ii) Assuming that the mixing height, h, is 500 m, and the box is 1000 m long, what is the concentration of SO_2? Give your answer in units of g m^{-3}.

  (iii) This concentration calculated in (ii) can be converted to a concentration in units of ppmv. Calculate the molar mass of SO_2 and hence the number of moles of SO_2 in each cubic metre.

  (iv) If one mole of SO_2 has a volume of 22.4 L, what volume does the SO_2 occupy in 1 m^3?

  (v)   1 m^3 contains 1000 L. To give your concentration of SO_2 in units of ppmv, multiply your answer to part (iv) by 1000 to give the volume occupied by SO_2 in 10^6 L (i.e., 1000 m^3).

The people most affected by air pollution are those situated downwind of the source. In the U.K. the prevailing winds are westerly to southwesterly, and hence, the better suburbs in U.K. towns are often situated on the west or southwest side of the town. A more detailed analysis of the frequency of wind direction can be obtained using a wind rose, as shown in Figure 4.16. The longer the line out from the centre in a particular direction, the more frequent are the winds *from* that direction and, therefore, the more that pollution is carried *away* from that direction.

Over long distances the trajectories of pollutant air masses can be quite curved as they follow wind circulation patterns. After waste gases leave the top of a chimney, the plume may rise to a considerable height before a horizontal travel begins (Figure 4.17). The major factor is the thermal buoyancy of the plume, and the hotter the gases, the greater the rise. A less important factor is the vertical momentum of the gases due to their exit velocity from the chimney top. Plume rise can effectively double the chimney height under low wind speeds, and so help achieve more effective dispersion. Conversely, any control technology for pollutant

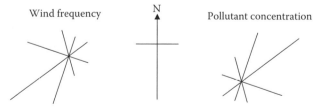

FIGURE 4.16 Effect of prevailing winds on airborne dispersal of pollutants: (a) wind frequency: the longer the line in any particular direction, the greater the frequency of winds *from* that direction; (b) Pollutant concentration: the longer the line in any direction, the greater the concentration of pollutants in that direction.

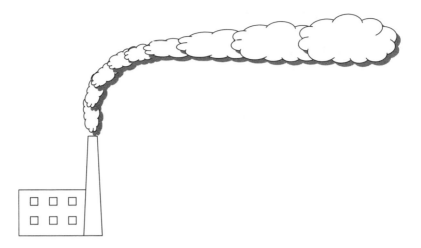

FIGURE 4.17 Emissions of plumes from chimneys.

reduction, which also reduces flue gas temperature (such as a scrubber) results in poorer dispersion of the plume by preventing it from rising to high altitudes. Also, thermal layering may affect the dispersion of a plume. The basic models of plume dispersion suggest that the maximum ground-level concentration will depend on the square of the chimney height, so that a 40% increase in chimney height should roughly halve the ground-level impact. Figure 4.18 shows the processes that may be involved between the emission of an air pollutant and its ultimate deposition on the ground.

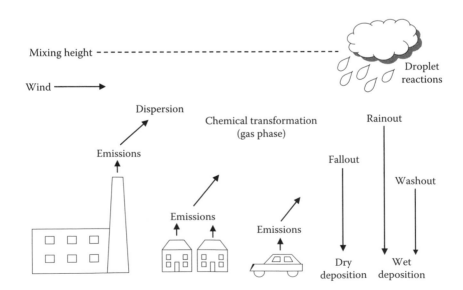

FIGURE 4.18 Emission processes.

4.6.4 INDOOR AIR QUALITY

Because the majority of people in industrialised countries spend a large proportion of their time indoors, whether at home, work, or leisure, indoor air quality is an important issue. Major pollutants include cigarette smoke, carbon monoxide, VOCs, and radon gas, and we will consider each of these categories of pollutant.

Cigarette smoke is a complex mixture of several toxic compounds including nitrosamines and polyaromatics (PAHs). The total number of components in smoke is large, particularly if trace amounts are included; according to a recent estimate the number of components is between 450 and 500 in the vapour phase and 3500 in the particulate phase. The introduction of filter tips has helped to reduce the toxic agents inhaled by smokers, but they are not 100% efficient and do not control the harmful constituents of side-smoke (i.e., smoke emitted between puffs). It has been shown that the toxic compounds in the side-smoke far exceed the quantity in the main-smoke. Tobacco smoke is an aerosol of volatile components in the vapour phase, and of semi- and nonvolatile compounds of particulate matter commonly called **tar**. Government legislation in many countries and states is being, or has been, introduced to restrict the number of public places where people can smoke, as it is a major heath risk to everyone within an indoor environment.

VOCs that affect indoor air quality (domestic and industrial) include solvents used in many chemical processes as well as in paints, varnishes, glues, surface coatings, cleaning materials, paint strippers, and nail varnish removers. VOCs are also produced as gases by domestic boilers and central heating systems, foam blowing gases, and breakdown products of polyurethane foam used in insulation (Figure 4.19). Many of these compounds are harmful or irritants, and some are toxic at greater or lesser levels. For many pollutants, maximum permitted occupational exposure limits (OELs) have been set. Types of OELs include 8 hour OELs, to set limits to exposure over a typical working day, as well as short-timescale OEL to limit short but high exposures that could be hidden when averaged over an 8-hour period. Some examples of workplace exposure limits (WELs) for a variety of compounds in the U.K. are given in Table 4.6.

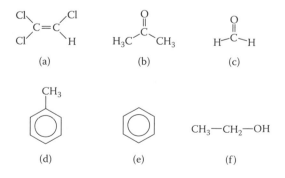

FIGURE 4.19 Examples of volatile organic compounds (VOCs). Example VOCs are (a) trichloroethylene (trichloroethene), (b) acetone (propanone), (c) formaldehyde (methanal), (d) toluene (methylbenzene), (e) benzene (f) ethanol.

TABLE 4.6
Example U.K. Workplace Exposure Limits

Substance	8-Hour TWA		15-Minute TWA	
	/ppmv	/mg m^{-3}	/ppm	/mg m^{-3}
Acetone	500	1210	1500	3620
Benzene	1	—	—	—
Ethanol	1000	1920	—	—
Formaldehyde	2	2.5	2	2.5
Toluene	50	191	150	574
CO	30	35	200	232
N$_2$O	100	183	—	—
Grain dust	—	10	—	—
Hardwood dust	—	5	—	—

Carbon monoxide (CO) is a colourless, odourless poisonous gas produced indoors principally from leaking vented combustion appliances (e.g., gas fires, water heaters, and central heating boilers), unvented combustion appliances (such as portable calor gas fires) and vehicles in poorly vented garages. The recommended maximum levels for indoors are 3 ppm above outdoor level (alert level) and 9 ppm (health level). The alert level is an indication of abnormal indoor concentration, and investigation of possible sources is advisable. The health level is the level at which persons with coronary artery disease would be affected.

Radon is a heavy, colourless, odourless radioactive gas produced by the radioactive decay of uranium-238 found mainly in granite rocks (see the ^{238}U radioactive decay sequence in Chapter 1, Figure 1.7). The main area in the U.K. where radon is a problem is in the southwest, which lies above an extensive area of granite rocks. When radon (in the form of the isotope ^{222}Rn) decays, it produces the radioactive isotope ^{218}Po, which in turn decays to radioactive ^{214}Pb by emitting an α particle. Whereas α radiation is stopped in a short distance in air, tiny particles of ^{218}Po can be inhaled and become lodged in the lungs, where they can cause cell damage because of their radioactivity. Because of its density, radon was found mainly in cellars until the advent of well-insulated, double-glazed houses, after which it was also detected elsewhere in houses. In such situations it can reach dangerous levels, up to 200 Bq m^{-3} (becquerels per cubic metre) in ground floor rooms, unless remedial action is taken. Remedial action might include ventilation bricks being installed low down near the floor, installation of ventilation fans, or sealing of the ground on which the home is built. In the U.K. a free radon-metering system is available from the National Radiological Protection Board (NRPB). The monitoring process takes 3 months to complete.

4.7 SUMMARY

In this chapter you should have learnt that:

Earth's atmosphere is divided vertically into layers, which are defined by changes in atmospheric temperature. Most air pollution stays in the lowest layer, the troposphere.

Earth quickly lost its primary atmosphere of hydrogen and helium. This was replaced by Earth's secondary atmosphere, formed by outgassing from the mantle, which was originally composed mostly of CO_2 with a few percent N_2 and virtually no O_2. Atmospheric O_2 levels began to rise markedly some 500 million years ago.

Essential elements such as carbon, nitrogen, and sulfur cycle through the atmosphere, hydrosphere, geosphere, and biosphere in what are termed biogeochemical cycles.

Anthropogenic activities (burning of fossil fuels and deforestation) have increased the rate at which CO_2 is being added to the atmosphere. Some of the excess CO_2 is being absorbed by the oceans.

The level of atmospheric CO_2 is increasing by about 1–2 ppmv per year because the rate of addition of CO_2 to the atmosphere now exceeds the rate at which CO_2 is being removed from the atmosphere.

The greenhouse effect keeps Earth warmer than it would be if the temperature were determined solely by the amount of solar radiation landing on Earth's surface. This is due to the absorption by greenhouse gases in the atmosphere of long-wavelength radiation that is re-emitted by Earth.

Global warming is occurring because of an enhanced greenhouse effect due to increasing levels of greenhouse gases in the atmosphere, in particular, CO_2.

Ozone in the ozone layer protects Earth from short-wavelength UV radiation. Chlorofluorocarbons and related groups of molecules are causing depletion of the ozone layer.

Air pollution may cause problems on a local, regional, or global scale. The severest effects are often felt downwind of the source of pollution.

Key pollutants causing air pollution include SO_2 and NO_x, which give rise to acid rain. Combustion of fossil fuels can lead to the formation of either classical smog or photochemical smog.

Answers to Self-Assessment Questions

CHAPTER 1

Q1.1   (i) $^{11}_{5}B$, boron

      (ii) $^{20}_{10}Ne$, neon

      (iii) $^{23}_{11}Na$, sodium

      (iv) $^{40}_{20}Ca$, calcium

      (v) $^{75}_{33}As$, arsenic

      (vi) $^{89}_{39}Y$, yttrium

      (vii) $^{103}_{45}Rh$, rhodium

      (viii) $^{181}_{73}Ta$, tantalum

      (ix) $^{209}_{84}Po$, polonium

Q1.2   (i)   These are some of the isotopes of tin ($Z = 50$): ^{117}Sn, ^{118}Sn, and ^{119}Sn

      (ii)   These are isotopes of silicon ($Z = 14$): ^{28}Si, ^{29}Si, and ^{30}Si

      (iii) These are isotopes of uranium ($Z = 92$): ^{236}U, ^{237}U, and ^{238}U.

Q1.3   To calculate the number of neutrons, use $N = A - Z$.

      (i)   For the three isotopes of tin listed, $N = 67, 68$, and 69, respectively.

      (ii)   For the three isotopes of silicon listed, $N = 14, 15$, and 16, respectively.

      (iii) For the three isotopes of uranium listed, $N = 144, 145$, and 146, respectively.

Q1.4   The atomic mass of copper is calculated as follows:

atomic mass $= (62.93 \times 69.5/100) + (64.93 \times 30.5/100)$ amu $= 63.54$ amu.

Q1.5   (i) $1s^2\ 2s^2$ *or* [He] $2s^2$

      (ii) $1s^2\ 2s^2\ 2p^6\ 3s^2\ 3p^2$ *or* [Ne] $3s^2\ 3p^2$

      (iii) $1s^2\ 2s^2\ 2p^6\ 3s^2\ 3p^6\ 4s^2$ *or* [Ar] $4s^2$

      (iv) $1s^2\ 2s^2\ 2p^6\ 3s^2\ 3p^6\ 4s^2\ 3d^{10}\ 4p^3$ *or* [Ar] $4s^2\ 3d^{10}\ 4p^3$

      (v) $1s^2\ 2s^2\ 2p^6\ 3s^2\ 3p^6\ 4s^2\ 3d^{10}\ 4p^6\ 5s^2\ 4d^{10}\ 5p^2$ *or* [Kr] $5s^2\ 4d^{10}\ 5p^2$

Q1.6   (i) $^{60}_{28}Ni$

      (ii) $^{11}_{5}B$

      (iii) $^{222}_{86}Rn$

      (iv) $^{7}_{3}Li$

Q1.7   (i)   Thirty years is one half-life of ^{137}Cs, so there will be half the original amount; i.e., there will be 40 g remaining.

(ii)   Sixty years is two half-lives of ^{137}Cs, so there will be $1/2 \times 1/2$ the original amount remaining; i.e., 1/4 of the original amount, which is 20 g. To put it another way, after another 30 years (the second half-life), there will be half the amount that there was after the first half-life, i.e., half of 40 g, which is 20 g.

(iii)   Ninety years is three half-lives of ^{137}Cs, so there will be $1/2 \times 1/2 \times 1/2$ the original amount remaining, i.e., 1/8 of the original amount, which is 10 g. To put it another way, after another 30 years (the third half-life), there will be half the amount that there were after the second half-life, i.e., half of 20 g, which is 10 g.

Q1.8   If the temperature, T, increases, then the value of nRT will also increase (we are considering a fixed number of molecules of gas). The volume V must therefore also increase (if p stays constant), so that the value of pV stays equal to nRT.

Q1.9   (i)   The stable state of CO_2 at 5.11 atm and $-100°C$ is as a solid.

        (ii)   The stable state of CO_2 at 5.11 atm and 25°C is as a gas.

        (iii)   The stable state of CO_2 at 67 atm and 0°C is as a liquid.

Q1.10   (i)   There are two N atoms, three O atoms, and four H atoms in NH_4NO_3, so the RMM of NH_4NO_3 = $(2 \times 14.01) + (3 \times 16.00) + (4 \times 1.01) = 70.06$

        (ii)   The molar mass of NH_4NO_3 is 70.06 g mol^{-1}.

Q1.11   (i)   The molar mass of NH_4NO_3 is 70.06 g mol^{-1}. The number of moles is calculated as below. Considering the units in the equation, g above and below the line cancel out, and mol^{-1} below the line becomes mol above the line.

$$\text{number of moles} = \frac{\text{mass of substance}}{\text{molar mass}} = \frac{140.12 \text{ g}}{70.06 \text{ g mol}^{-1}} = 2 \text{ mol}$$

        (ii) $$\text{number of moles} = \frac{3.503 \text{ g}}{70.06 \text{ g mol}^{-1}} = 0.05 \text{ mol}$$

Q1.12   (i)   A sensible choice would be to use km (10^3 m), giving 12.86 km.

        (ii)   You could use MJ (10^6 J), giving 6.83 MJ, or kJ (10^3 J), giving 6830 kJ.

        (iii)   The best choice would be μg (10^{-6} g), giving 3.74 μg.

        (iv)   You could use mL (10^{-3} L), giving 29.2 mL.

Q1.13   The molar mass of magnesium sulfate = $24.31 + 32.07 + (4 \times 16.00)$ g mol^{-1} = 120.38 g mol^{-1}. The concentration of the solution is calculated as follows:

$$\text{concentration} = \frac{\text{mass} \times 1000}{\text{molar mass} \times \text{volume}} = \frac{5.32 \times 1000}{120.38 \times 100} M = 0.442 \ M$$

Q1.14 (i)   Hydrogen, with an electronic configuration of $1s^1$, has one valence electron, and therefore needs to share one more for a stable arrangement. The electronic configuration of chlorine is [Ne] $3s^2$ $3p^5$. It has seven valence electrons, and needs one more to complete the stable octet. Hydrogen therefore shares its one electron with chlorine. The Lewis diagram for HCl is shown below in Figure A1.

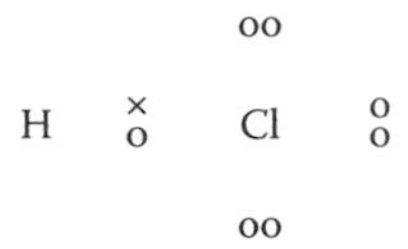

FIGURE A1 Lewis diagram of hydrogen chloride, HCl.

(ii)   The electronic configuration of oxygen is [He] $2s^2$ $2p^4$. Each oxygen has six valence electrons, and needs to share another two electrons to achieve a stable shell. Therefore, each oxygen shares two electrons, giving an O=O double bond. The Lewis structure of O_2 is shown below in Figure A2.

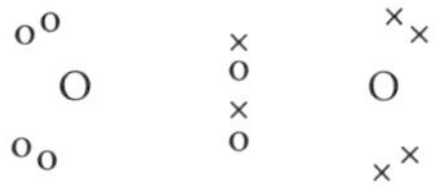

FIGURE A2 Lewis diagram of oxygen, O_2.

Q1.15   The electronic configuration of nitrogen is [He] $2s^2$ $2p^3$. It has five valence electrons, and needs to share another three electrons. Therefore, each nitrogen shares three electrons, giving an N≡N triple bond. The Lewis structure of N_2 is shown in Figure A3.

FIGURE A3 Lewis diagram of nitrogen, N_2.

Q1.16   In NO_2, the nitrogen atom sits between the two oxygen atoms. There are, in fact, two possible Lewis structures shown in Figure A4. In structure (a) both oxygens have stable octets, and the nitrogen has an unpaired electron, whereas in structure (b) the nitrogen and the left-hand oxygen have a stable octet and the right-hand oxygen has an unpaired electron. This molecule readily accepts an electron to form the NO_2^- anion, which completes the stable octet on the third atom (N in (a) and the right-hand O in (b)).

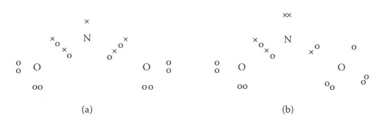

(a)   (b)

FIGURE A4 Lewis diagrams of NO_2.

Q1.17 Two d orbitals can overlap as shown in Figure A5, forming a π bond (the electron density lies above and below the plane of the atoms).

FIGURE A5 Overlap of two d orbitals forming a π bond.

Q1.18 Two p orbitals can overlap end on as shown in Figure A6 to form a σ bond (the electron density lies between the atoms).

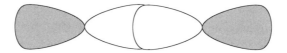

FIGURE A6 Overlap of p orbitals forming a σ bond.

Q1.19 One way in which the overlap of an s and a d orbital will *not* give a bond is shown in Figure A7.

FIGURE A7 Overlap of an s orbital and a d orbital in which a bond is not formed.

Q1.20 The electronic configuration of carbon is [He] $2s^2 2p^2$. Carbon therefore needs four electrons to make a stable octet. It can achieve this by sharing four electrons from the three oxygen atoms that are shown in Figure A8 as one C=O and two C-O bonds. The two added electrons are shown as filled circles, giving the overall –2 charge.

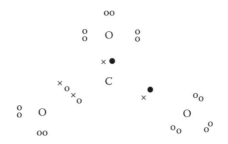

FIGURE A8 Lewis diagram of carbonate, CO_3^{2-}.

Q1.21   KI contains K^+ and I^- ions.

Q1.22   The +2 charge on the Ca^{2+} cation is balanced by the –2 charge on the SO_4^{2-} anion, so the formula of anhydrite is $CaSO_4$.

Q1.23   Two OH^- anions are needed to balance every one Mg^{2+} cation, so the formula of brucite is $Mg(OH)_2$.

Q1.24   (i)   There are three different structural isomers for C_3H_8O. Their structures are shown in Figure A9 together with their condensed formulas.

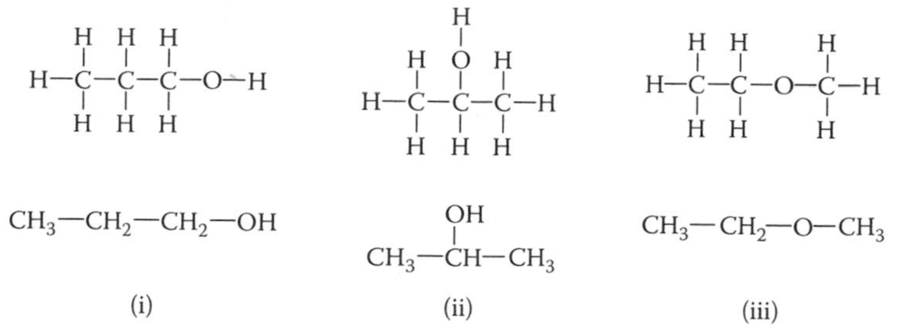

$$CH_3-CH_2-CH_2-OH$$

(i)

$$CH_3-\overset{\displaystyle OH}{\underset{\displaystyle |}{CH}}-CH_3$$

(ii)

$$CH_3-CH_2-O-CH_3$$

(iii)

FIGURE A9 Structural isomers of C_3H_8O.

(ii)   There are four structural isomers for C_3H_9N. Their structures are
        shown in Figure A10 together with their condensed formulas.

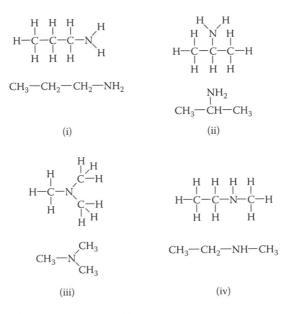

FIGURE A10   Structural isomers of C_3H_9N.

Q1.25 Because there is only one unpaired electron on the nitrogen, the repul-
       sion between this one electron and the electrons in the N-O bonds will
       be less than the repulsion between the electrons in the bonds. Therefore,
       the two oxygen atoms will tend to push further apart, opening the angle
       between the oxygens and closing the angles between the lone electron
       and the oxygen atoms.

Q1.26 There are four carbon atoms in butane and, therefore, four molecules
       of carbon dioxide, CO_2, will be formed. There are ten atoms of hydrogen
       in butane, so five molecules of water, H_2O, will be formed. This means
       that the number of oxygen atoms required to balance the equation will
       be 4×2 (for CO_2) plus 5×1 (for H_2O) = 13. Therefore, 13/2 molecules
       of oxygen, O_2, will be required, and the balanced reaction will be:

$$C_4H_{10} + 13/2O_2 \rightarrow 4CO_2 + 5H_2O$$

Q1.27 (i)   The structural formula of octane is shown in Figure A11. There
           are 7 C–C bonds and 18 C–H bonds that will be broken when
           petrol is burned.

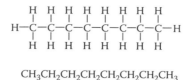

$$CH_3CH_2CH_2CH_2CH_2CH_2CH_2CH_3$$

FIGURE A11 The structural and condensed formulas of octane.

    (ii) 25/2 O=O bonds will be broken. 16 C=O bonds (two per molecule) and 18 H–O bonds (two per molecule) will be made.

   (iii) The bond energies of C-C, C-H, O=O, C=O, and H-O are 347, 414, 498, 803, and 464 kJ mol^1, respectively.

The energy taken in on breaking bonds =
$(7 \times 347) + (18 \times 414) + (25 \times 498/2)$ kJ mol^{-1} = 16,100 kJ mol^{-1}

The energy given out in making bonds =
$(16 \times 803) + (18 \times 464)$ kJ mol^{-1} = 21200 kJ mol^{-1}

The enthalpy change = 16,100 – 21,200 kJ mol^{-1} = 5100 kJ mol^{-1}

Q1.28 (i) The equilibrium constant is given by the expression

$$K = \frac{[SO_3]^2}{[SO_2]^2 \times [O_2]}$$

    (ii) As the volume of the flask is 1 L, we can take the number of moles formed as being the concentration.

$$K = \frac{0.895^2}{0.063^2 \times 0.721} \text{ mol L}^{-1} = 2.80 \times 10^2 \text{ mol L}^{-1}$$

Q1.29 (i) The atomic number is 16 and, therefore, the element is sulfur, S.

    (ii) $^{32}_{16}$S will have 16 protons, 16 electrons, and 32 – 16 neutrons, i.e., 16 neutrons. $^{33}_{16}$S will have 16 protons, 16 electrons, and 33 – 16 neutrons, i.e., 17 neutrons.

Q1.30 (i) The ions are Ca^{2+}, Mg^{2+}, and CO_3^{2-}.

    (ii) The RMM of dolomite = 40.01 + 24.31 + (2 × 12.01) + (6 × 16.00) = 184.34

   (iii) The molar mass of dolomite therefore is 184.34 g mol^1.

   (iv) The mass in kg must be multiplied by 1000 to convert it to a mass in g.

$$\text{Number of moles} = \frac{\text{mass in g}}{\text{molar mass}} = \frac{2.5 \times 1000}{184.34} \text{ mol} = 13.56 \text{ mol}$$

Q1.31 (i)   The ions are Na^+ and SO_4^{2-}.
       (ii)   We must first calculate the molar mass of Na_2SO_4.

Molar mass $= (2 \times 22.99) + 32.07 + (4 \times 16.00)$ g mol^{-1} = 142.05 g mol^{-1}.

$$\text{Concentration} = \frac{\text{mass} \times 1000}{\text{molar mass} \times \text{volume of solution}} = \frac{10.3 \times 1000}{142.05 \times 250} \ M = 0.290 \ M$$

   (iii)   There are two Na^+ in the formula and, therefore, there are two
          moles of Na^+ for every mole of Na_2SO_4. Therefore, the concen-
          tration of Na^+ is 0.580 M (i.e., 0.580 mol L^{-1}).
   (iv)   To convert concentration in mol L^{-1} to mg L^{-1}, first convert to g
          L^{-1} by multiplying by the molar mass of Na.

Concentration Na^+ = 0.580 mol L^{-1} = 0.580 × 22.99 g L^{-1} = 13.3 g mol^{-1}.

To convert g L^{-1} to mg L^{-1}, multiply by 1000.

Concentration Na^+ = 13.3 × 1000 mg L^{-1} = 1.33 × 10^4 mg L^{-1}

Q1.32 (i)   There is an –O–H group, therefore this is an alcohol.
      (ii)   There is a –C≡N group, therefore this is a nitrile.
      (iii)   There is a –CO_2^- group, therefore this is an ester.
      (iv)   There is a –Br atom, therefore this is a bromoalkane.
      (v)   There are no functional groups. This is an alkane.
      (vi)   There is a –NO_2 group and an aromatic ring, therefore this is an
             aromatic nitro compound.
Q1.33 (i)   The equilibrium constant is given by the following equation:

$$K = \frac{[NO]^2}{[N_2] \times [O_2]}$$

   (ii)   Rewriting this in terms of partial pressures gives us:

$$K = \frac{p_{NO}^2}{p_{N_2} \times p_{O_2}}$$

   (iii)   Rearranging the equation in (ii) gives:

$$p_{NO} = \sqrt{K \times p_{N_2} \times p_{O_2}} = \sqrt{1.0 \times 10^{-5} \times 0.78 \times 0.21} \text{ atm} = 1.3 \times 10^{-3} \text{ atm}$$

CHAPTER 2

Q2.1 A rock with 70% SiO_2 will be classed as a felsic rock.

Q2.2 The approximate composition (measuring from Chapter 2, Figure 2.16) is 10% potassium feldspar, 12% quartz, 49% plagioclase, 11% biotite and 18% amphibole.

Q2.3 Generally speaking, the *higher* the temperature of crystallisation of the minerals forming a particular igneous rock, the *lower* the percentage of SiO_2 in that rock. For example olivine crystallises at high temperature and is found in ultramafic and mafic rocks with less than 52% SiO_2.

Q2.4 An igneous rock with 50% SiO_2 is classed as a mafic rock. As olivine and pyroxene crystallise out, the SiO_2 content of the residual magma will increase.

Q2.5 If the originally formed crystals are separated from the magma when the SiO_2 content of the magma has increased to 55%, then an intermediate rock will form from the crystallisation of the residual magma.

Q2.6 Reading from Figure A12:

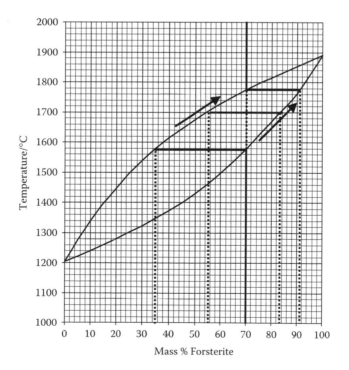

FIGURE A12 Phase diagram of the forsterite–fayalite system.

  (i) Melting starts at 1575°C.

  (ii) The composition of the first melt is $Fo_{35}Fa_{65}$.

  (iii) The last crystals melt at 1775°C.

  (iv) The composition of the last crystals to melt is $Fo_{91}Fa_9$.

  (v) The composition of the final melt is the same as the composition of the original solid, which is $Fo_{70}Fa_{30}$.

  (vi) The distance a from the isopleth to the liquidus = 70 − 55 = 15

The distance b from the solidus to the isopleth = 83 − 70 = 13

$$\text{percentage of solid} = \frac{a}{a+b} \times 100\% = \frac{15}{28} \times 100\% = 54\%$$

$$\text{percentage of liquid} = \frac{b}{a+b} \times 100\% = \frac{13}{28} \times 100\% = 46\%$$

Q2.7 Reading from Figure A13

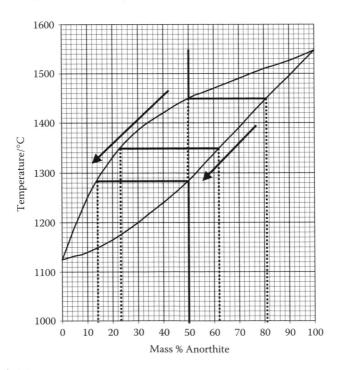

FIGURE A13 Phase diagram of the anorthite–albite system.

  (i) Crystallisation starts at 1450°C.

  (ii) The composition of the first crystals to form is $An_{81}Ab_{19}$.

(iii) At 1350°C the composition of the melt will be $An_{23}Ab_{77}$, and the composition of the solid will be $An_{62}Ab_{38}$.

(iv) The distance a from the isopleth to the liquidus = 50 − 23 = 27.

The distance b from the solidus to the isopleth = 62 − 50 = 12.

$$\text{percentage of solid} = \frac{a}{a+b} \times 100\% = \frac{27}{39} \times 100\% = 69\%$$

$$\text{percentage of liquid} = \frac{b}{a+b} \times 100\% = \frac{12}{39} \times 100\% = 31\%$$

(v) Crystallisation will be complete at 1285°C.

(vi) The composition of the last drop of melt will be $An_{14}Ab_{86}$, and the composition of the solid will be the same as the original melt, which is $An_{50}Ab_{50}$.

Q2.8 Reading from Figure A14:

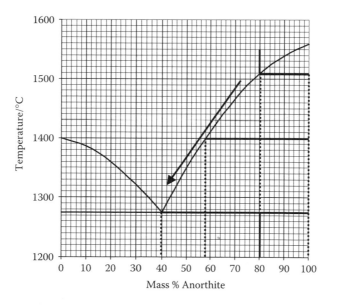

FIGURE A14 Phase diagram of the anorthite–diopside system.

(i) Crystallisation will begin at 1510°C.

(ii) The first crystals to form will have a composition of $An_{100}Di_0$; i.e., they will be pure anorthite.

(iii) At 1400°C the composition of the melt will be $An_{58}Di_{42}$, and the composition of the solid will be $An_{100}Di_0$.

(iv) As cooling continues, there will be a decreasing proportion of anorthite in the melt.

(v) At 1276°C the composition of the melt will have reached $An_{40}Di_{60}$. Crystallisation will continue at this temperature until all the remaining melt has crystallised.

Q2.9 Reading from Figure A15:

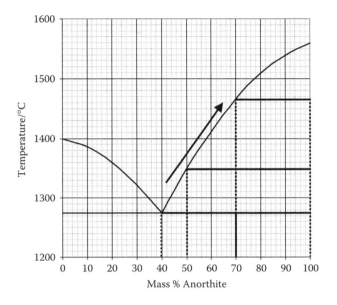

FIGURE A15 Phase diagram of the anorthite–diopside system.

(i) Melting will start at 1276°C.

(ii) The composition of the melt will be $An_{40}Di_{60}$.

(iii) The melt changes composition when all the diopside is used up. At this point the temperature will start to rise.

(iv) At 1350°C, the melt will have a composition of $An_{50}Di_{50}$ and the residual solid will be pure anorthite.

(v) Melting will be complete at 1465°C.

(vi) At 1465°C, the melt will have the same composition as the original solid, which is $An_{70}Di_{30}$, and the last trace of solid will be pure anorthite, $An_{100}Di_0$.

Q2.10 (i) The ionic radius of Sr^{2+} is very large (132 pm) compared to Mg^{2+} (86 pm) or Fe^{2+} (92 pm), so it would distort the olivine crystal lattice. It therefore preferentially stays in the melt and is incompatible.

(ii) The ionic radius of Sr^{2+} is larger than that of Ca^{2+} (114 pm) but not so large that it cannot squeeze into the plagioclase lattice in preference to staying in the melt. It is therefore compatible. (Remember that K^+, with a radius of 152 pm (rather larger than Sr^{2+}), can only replace Na^+ at high temperatures, when the lattice has expanded.)

Q2.11 (i)   A rock that is 70% clay and 30% calcite is a calcareous mudstone.

(ii) A rock that is 60% calcite and 40% quartz is sandy limestone.

(iii) A rock that is 65% quartz and 35% clay is muddy sandstone.

Q2.12 Andalusite is the stable polymorph at 200 MPa and 550°C.

Q2.13 Kyanite and sillimanite will be in equilbrium at 612°C when at a pressure of 600 MPa.

Q2.14 (i)   Muscovite and quartz will be the stable assemblage at 500 MPa and 500°C.

(ii) Feldspar and sillimanite Al_2SiO_5 will be the stable assemblage.

Q2.15 (i)   Muscovite and quartz will be in equilibrium with feldspar and Al_2SiO_5 at 723°C.

(ii) At these conditions, the stable polymorph of Al_2SiO_5 will be sillimanite.

Q2.16 The formula of olivine is $(Mg,Fe)_2SiO_4$ and, therefore, there should be a total of 2 moles of Mg+Fe in the formula. In this particular olivine, therefore, there will be 1.4 moles of Fe^{2+}, and the molecular formula will be $Mg_{0.6}Fe_{1.4}SiO_4$.

Q2.17 The formula of orthopyroxene is $(Mg,Fe)_2Si_2O_6$ and, therefore, there should be a total of 2 moles of Mg+Fe in the formula. In this particular orthopyroxene, therefore, there will be 0.7 moles of Mg^{2+}, and the molecular formula will be $Mg_{0.7}Fe_{1.3}Si_2O_6$.

Q2.18 In order to maintain charge balance when Al^{3+} replaces Si^{4+}, another +1 charge needs to be added to the formula. In (i), K^+ has been added, in (ii) Ca^{2+} has been added, and in (iii) Al^{3+} has been added. The correct answer therefore is (i).

Q2.19 A mafic igneous rock with 48% SiO_2 has about 8% olivine, 58% pyroxene, and 34% plagioclase.

Q2.20 Rb^+ is larger than either Na^+ (116 pm), which occurs in plagioclase feldspar, or K^+ (152 pm), which occurs in potassium feldspar. It is, however, much closer in size to K^+ and, therefore, you would expect it to be more compatible in potassium feldspar than plagioclase. This is indeed the case. K_d for Rb^+ in potassium feldspar is 0.7; i.e. it is only slightly incompatible, whereas it is 0.1 in plagioclase.

Q2.21 Yttrium, Y^{3+}, and ytterbium, Yb^{3+}, despite being larger than scandium, will replace Al^{3+} in garnet and, therefore, scandium, as you should have expected, is highly compatible in garnet, having a K_d value of 23.

CHAPTER 3

Q3.1 H_2SO_4 can act as a dibasic acid by giving up two protons:

$$H_2SO_4 \rightleftarrows H^+ + HSO_4^- \rightleftarrows 2H^+ + SO_4^{2-}$$

Q3.2 $H_2PO_4^-$, and HCO_3^- are amphibasic (amphiprotic). H_3PO_4 can only lose protons and, therefore, it can only act as an acid, whereas CO_3^{2-} can only gain protons and, therefore, it can only act as a base.

   (ii) $H_3PO_4 \rightleftarrows H^+ + H_2PO_4^- \rightleftarrows 2H^+ + HPO_4^{2-}$

   (iii) $H_2CO_3 \rightleftarrows H^+ + HCO_3^- \rightleftarrows 2H^+ + CO_3^{2-}$

Q3.3   (i) NO_3^-

   (ii) Br^-

   (iii) S^{2-}

Q3.4   (i)   HCN

   (ii) $HClO_4$

   (iii) H_2S

Q3.5   The conjugate base of H_2SO_4 is HSO_4^-. The conjugate acid of NH_3 is NH_4^+. As H_2SO_4 is a stronger acid than NH_4^+, the reaction will go to the right.

$$H_2SO_4 + NH_3 \rightarrow HSO_4^- + NH_4^+$$

Q3.6   The conjugate base of H_2CO_3 is HCO_3^-. The conjugate acid of SO_4^{2-} is HSO_4^-. As HSO_4^- is a stronger acid than H_2CO_3, the reaction will go to the left.

$$H_2CO_3 + SO_4^{2-} \leftarrow HCO_3^- + HSO_4^-$$

Q3.7   (i)   The concentration of H_3O^+ cations formed will be the same as that of A^- anions formed, so $[H_3O^+] = 0.005\ M$.

   (ii) $[HA] = 0.2 - [A] = 0.2 - 0.005\ M = 0.195\ M$

   (iii)   % ionisation $= \dfrac{[A^-]}{[HA]} \times 100\% = \dfrac{0.005}{0.195} \times 100\% = 2.56\%$

   (iv) $K_a = \dfrac{[H_3O^+] \times [A^-]}{[HA]} = \dfrac{0.005 \times 0.005}{0.195} 1.28 \times 10^{-4}$

Q3.8   Assume that [HA] will hardly change on ionisation, because the degree of ionisation will be very small, so $[HA] = 0.05\ M$. $[H_3O^+] = [A^-]$, and therefore $[H_3O^+] \times [A] = [A]^2$.

$$K_a = 1.28 \times 10^{-4} = \frac{[A^-] \times [H_3O^+]}{[HA]} = \frac{[A^-]^2}{0.05}$$

$$[A^-]^2 = 1.28 \times 10^{-4} \times 0.05 = 6.4 \times 10^{-6}$$

$$[A^-] = \sqrt{6.4 \times 10^{-6}}\ \text{mol L}^{-1} = 0.0025\ \text{mol L}^{-1}$$

Q3.9 $pK_a = -\log_{10} K_a = -\log_{10} (4.2 \times 10^{-7}) = 6.38$

Q3.10 $K_w = [H_3O^+] \times [OH^-]$

$$[OH^-] = \frac{K_w}{[H_3O^+]} = \frac{1.0 \times 10^{-14}}{4.5 \times 10^{-5}} M = 2.22 \times 10^{-10} M$$

Q3.11 (i) If there are 0.0040 moles of HCl in 250 mL, then there will be four times as many moles of HCl in 1000 mL (i.e., 1 L).

$$[HCl] = \frac{0.0040 \times 1000}{250} \text{ mol L}^{-1} = 0.016 \text{ mol L}^{-1}$$

(ii) The concentration of H^+ will be the same as the concentration of HCl, i.e., 0.016 mol L^{-1}.

(iii) $[OH^-] = \dfrac{K_w}{[H^+]} = \dfrac{1.0 \times 10^{-14}}{0.016}$ mol L$^{-1} = 6.25 \times 10^{-13}$ mol L^{-1}

Q3.12 $K_w = K_a \times K_b$

$$K_b = \frac{K_w}{K_a} = \frac{1.0 \times 10^{-14}}{4.2 \times 10^{-7}} = 2.4 \times 10^{-8}$$

Q3.13 $pH = -\log_{10} [H^+] = -\log_{10} (1 \times 10^{-7}) = -(-7) = 7$
Note that the concentration is given in the question as 10^{-7} mol L^{-1}. This is shorthand for 1×10^{-7} mol L^{-1}.

Q3.14 $pH = -\log_{10} [H^+] = 8.1$, therefore $\log_{10} [H+] = -8.1$

$$[H+] = \text{antilog}_{10} (-8.1) = 7.9 \times 10^{-9} \text{ mol L}^{-1}$$

$$K_w = [H^+] \times [OH^-]$$

$$[OH^-] = \frac{K_w}{[H^+]} = \frac{1.0 \times 10^{-14}}{7.9 \times 10^{-9}} \text{ mol L}^{-1} = 1.27 \times 10^{-6} \text{ mol L}^{-1}$$

Q3.15 $pK_a = -\log_{10} K_a = -\log_{10} (4.2 \times 10^{-7}) = 6.4$

Q3.16 First, calculate the number of moles of NaOH that were used.

Moles of NaOH –

$$\frac{[NaOH] \times \text{volume of NaOH}}{1000} = \frac{0.0500 \times 22.7}{1000} \text{ mol} = 1.13 \times 10^{-3} \text{ mol}$$

The reaction between NaOH and ethanoic acid is:

$$NaOH + CH_3CO_2H \rightarrow CH_3CO_2Na + H_2O$$

One mole of NaOH is required for every one mole of ethanoic acid, and therefore there are 1.13×10^{-3} moles of ethanoic acid. This number of moles is in 50 mL, and therefore the concentration of ethanoic acid, $[CH_3CO_2H]$, will be:

$$[CH_3CO_2H] = \frac{\text{moles of } CH_3CO_2H \times 1000}{\text{volume of } CH_3CO_2H} = \frac{1.13 \times 10^{-3} \times 1000}{50} = 0.0227 \text{ mol L}^{-1}$$

Q3.17　The reaction between NaOH and H_2SO_4 is:

$$2NaOH + H_2SO_4 \rightarrow Na_2SO_4 + 2H_2O$$

Two moles of NaOH are required for every one mole of H_2SO_4 and, therefore, 0.4 moles of NaOH will be required to neutralise 0.2 moles of H_2SO_4.

Q3.18　The pK_a of the system $= -\log(1.8 \times 10^{-5}) = 4.7$

$$pH = pK_a + \log\left(\frac{[\text{conjugate base}]}{[\text{acid}]}\right) = 4.7 + \log\frac{0.5}{1.0} = 4.7 + -0.3 = 4.4$$

Q3.19　The pK_a of the system $= -\log(6.2 \times 10^8) = 7.21$

$$pH = pK_a + \log\left(\frac{[HPO_4^{2-}]}{[H_2PO_4^-]}\right)$$

$$\log\left(\frac{[HPO_4^{2-}]}{[H_2PO_4^-]}\right) = pH - pK_a = 6.85 - 7.21 = -0.36$$

$$\frac{[HPO_4^{2-}]}{0.500} = \text{antilog}(-0.36) = 0.44$$

$$[HPO_4^{2-}] = 0.44 \times 0.500 \text{ mol L}^{-1} = 0.22 \text{ mol L}^{-1}$$

Q3.20　(i)　$K_{sp} = [Ba^{2+}] \times [SO_4^{2-}]$

(ii) $[Ba^{2+}] \times [SO_4^{2-}] = 0.05 \times 3.0 \times 10^{-6} = 1.5 \times 10^{-7}$. This is larger than 1×10^{-10} and, therefore, $BaSO_4$ will precipitate out of solution.

(iii) $[Ba^{2+}] \times [SO_4^{2-}] = 1.0 \times 10^{-6} \times 3.0 \times 10^{-6} = 3.0 \times 10^{-12}$. This is smaller than 1×10^{-10} and, therefore, $BaSO_4$ will not precipitate out of solution. Addition of 1 M H_2SO_4 will result in the precipitation of $BaSO_4$ as the K_{sp} will then be exceeded.

Q3.21 First calculate the concentration of dissolved CO_2, $[H_2CO_3]$ using Henry's Law:

$$[H_2CO_3] = K_H \times p_{CO_2} = 3.4 \times 10^{-2} \times 3750 \times 10^{-6} \text{ mol L}^{-1} = 1.28 \times 10^{-4} \text{ mol L}^{-1}$$

Then, using Equation 3.43, calculate $[H^+]$:

$$[H^+] = \sqrt{K_1 \times [H_2CO_3]} = \sqrt{4.26 \times 10^{-7} \times 1.28 \times 10^{-4}} \text{ mol L}^{-1} = 7.38 \times 10^{-6} \text{ mol L}^{-1}$$

Now you can calculate pH:

$$pH = -\log_{10} [H^+] = -\log_{10} (7.38 \times 10^{-6}) = 5.13$$

Therefore, the rain would be more acidic, by nearly half a pH unit, with higher levels of atmospheric CO_2.

Q3.22 This is exactly the same as Reaction 2.11:

$$CO_{2(aq)} + H_2O + CaCO_3 \rightarrow Ca^{2+} + 2HCO_3^-$$

Q3.23 (i) The oxidation state of H is +1 as usual; therefore, the oxidation state of S must be –2.

(ii) The oxidation state of O is –2 as usual; therefore, the oxidation state of N must be +5 (don't forget the –1 charge on the ion).

(iii) The oxidation state of H is +1 as usual, so the oxidation state of N is –3.

(iv) Ca^{2+} is a monatomic ion and therefore the oxidation state of Ca is the same as the charge, +2.

(v) The oxidation state of O is –2 as usual, so the oxidation state of Fe must be +3.

Q3.24 If the oxidation state of Fe is +2, then the oxidation state of each S must be –1.

Q3.25 The oxidation state of O is –2 as usual. There is an overall charge of –2 on the ion, so S must have an oxidation state of +6.

Q3.26 The gases coming out from the vents include sulfur in its most reduced state, of –2 in the form of H_2S. As the gases are vented into the atmosphere, H_2S will tend to oxidise. In acidic conditions, the oxidation of sulfur passes through the small region of stabililty of elemental sulfur, S and, hence, the yellow sulfur deposits are formed.

Q3.27 Sulfur is in the −1 oxidation state, which is nearly its lowest oxidation state, and iron is in its reduced state of +2; therefore, FeS$_2$ will be located near the bottom of the stability diagrams. See Figures 3.12 and 3.13.

Q3.28 (i) Water is in the solid state, i.e., is in the form of ice.

(ii) If we increase the pressure while keeping the temperature constant, then eventually the solid–water phase transition will be crossed, and the ice will melt. This behaviour is different from most materials in which an increase in pressure usually causes a liquid to solidify. It is the reason why ice skates work. The weight of a person on the narrow blade causes the pressure on the ice to increase sufficiently to melt the ice, and this lubricates the movement of the skate over the ice.

(iii) If we increase the temperature while keeping constant pressure then the solid-liquid phase transition will be crossed and the ice will melt. If we continue to increase the temperature, then the liquid–gas phase transition will be crossed, and the water will boil and turn to vapour.

Q3.29 We must first convert the water flow as a volume per second to mass per second. 1 g of water occupies 1 cm^3, so in 1 m^3 there are 100 × 100 × 100 g, i.e., there are 10^6 g. The water flow, therefore, is 60 × 10^{12} g s^{-1}.

The energy transported per second = 60 × 10^{12} × 4.18 × 4 J s^{-1} = 1.0 × 10^{15} J s^{-1}.

This is 3.6 × 10^{18} J hr^{-1}.

Q3.30 HCl is giving up H$^+$, so HCl is the acid and Cl$^-$ is its conjugate base. In solution Na$_2$CO$_3$ will ionise to form CO$_3^{2-}$. This has accepted a proton to form HCO$_3^-$, and therefore CO$_3^{2-}$ is the base and HCO$_3^-$ is its conjugate acid.

Q3.31 (i) pH = −log$_{10}$[H$^+$] and, therefore, [H$^+$] = 10^{-pH} = 10$^{-8.5}$ = 3.16 × 10^{-9}

(ii) A concentration of 136 mg L^{-1} is equivalent to 0.136 g L^{-1}, i.e. divide by 1000 to convert mg to g. To convert this to units of mol L^{-1}, divide by the molar mass of HCO$_3^-$, which is 61.02 g mol^{-1}. This gives [HCO$_3^-$] = 2.23×10^{-3} mol L^{-1}.

(iii) $K_{a2} = \dfrac{[CO_3^{2-}]\times[H^+]}{[HCO_3^-]}$

$$[CO_3^{2-}] = \frac{K_{a2}\times[HCO_3^-]}{[H^+]} = \frac{4.68\times10^{-11}\times2.23\times10^{-3}}{3.16\times10^{-9}}\ \text{mol L}^{-1} = 3.30\times10^{-5}\ \text{mol L}^{-1}$$

(iv) Alkalinity = $\Sigma([HCO_3^-] + 2 \times [CO_3^{-2}]) = 2.23 \times 10^{-3} + 2 \times (3.30 \times 10^{-5})$ mol L^{-1} = 2.30×10^{-3} mol L^{-1}

(v) The molar mass of $CaCO_3$ (to 2 decimal places) is 100.02 g mol^{-1}.

Alkalinity = $2.30 \times 10^{-3} \times 1000 \times 100.02/2$ mg L^{-1} $CaCO_3$ = 115 mg L^{-1} $CaCO_3$

Q3.32 (i) $K_{sp} = [Ca^{2+}] \times [CO_3^{2-}]$

$$[Ca^{2+}] = \frac{K_{sp}}{[CO_3^{2-}]} = \frac{3.8 \times 10^{-9}}{2.45 \times 10^{-5}} \text{ mol L}^{-1} = 1.55 \times 10^{-4} \text{ mol L}^{-1}$$

(ii) To convert units of mol L^{-1} to mg L^{-1}, multiply by 1000 × (molar mass of Ca^{2+}).

$[Ca^{2+}] = 1.55 \times 10^{-4} \times 1000 \times 40.01$ mg L^{-1} = 6.20 mg L^{-1}

Q3.33 First calculate the equivalent $[Ca^{2+}]$:

$$\text{Equivalent } [Ca^{2+}] = [Ca^{2+}] + \frac{[Mg^{2+}] \times RAMCa}{RAMMg}$$

$$= 28.5 + \frac{2.3 \times 40.01}{24.31} \text{ mg L}^{-1} = 32.3 \text{ mg L}^{-1}$$

$$\text{Hardness} = [Ca^{2+}] \times \frac{RMMCaCO_3}{RAMCa^{2+}}$$

$$= 32.3 \times \frac{100.02}{40.01} \text{ mg L}^{-1} \text{ } CaCO_3 = 80.7 \text{ mg L}^{-1} \text{ } CaCO_3$$

Q3.34 The oxidation state of hydrogen is +1, and the oxidation state of oxygen is −2, as usual. Na will be in the form of Na$^+$ ions and, therefore, its oxidation state is +1. the whole compound has no overall charge, so the oxidation states add up to 0. We can now work out the oxidation state of carbon. Let the oxidation state of carbon = x. Remember that the oxidation state of oxygen must be multiplied by 3 because there are three oxygens in the molecular formula:

$$0 = +1 + +1 + x + (3 \times -2)$$

$$0 = -4 + x$$

$$x = +4$$

The oxidation state of carbon is +4

CHAPTER 4

Q4.1   Oxygen is a very reactive gas. As the atmospheric levels of O_2 increase, the ease with which combustible substances including trees and other vegetation will catch fire (e.g., following a lightning strike) will increase. At sufficiently high levels of oxygen, spontaneous combustion may take place. All this will tend to limit the amount of oxygen in the atmosphere, as the oxygen will be used up in combustion.

Q4.2   (i)   The molar mass of octane $= 8 \times 12.01 + 18 \times 1.008$ g mol^{-1} = 114.22 g mol^{-1}.

$$\text{Moles octane} = \frac{\text{mass octane} \times 1000}{\text{molar mass octane}} = \frac{29.5 \times 1000}{114.22} \text{ mol} = 258.3 \text{ mol}$$

The factor of 1000 is included to convert kg to g.
(ii)   Number of moles of CO_2 formed $= 8 \times 258.3$ mol = 2066 mol
(iii)   The molar mass of $CO_2 = 12.01 + 2 \times 16.00$ g mol^{-1} = 44.01 g mol^{-1}

Mass CO_2 formed $= 44.01 \times 2066$ g $= 90900$ g $= 90.9$ kg

Q4.3   (i)   A wavelength of 255 nm $= 255 \times 10^{-9}$ m $= 2.55 \times 10^{-7}$ m
(ii)   The energy of this radiation is calculated as shown below:

$$E = \frac{h \times c \times N_A}{1000 \times \lambda} = \frac{6.626 \times 10^{-34} \times 2.998 \times 10^8 \times 6.022 \times 10^{23}}{1000 \times 2.55 \times 10^{-7}} \text{ kJ mol}^{-1} = 469 \text{ kJ mol}^{-1}$$

(iii)   Bonds that would be broken by radiation of this wavelength include H–H (436 kJ mol^{-1}), C–O (351 kJ mol^{-1}), and C–H (414 kJ mol^{-1}).

Q4.4   (i)   Rearranging Equation 4.16 gives

$$\lambda = \frac{h \times c \times N_A}{1000 \times E} = \frac{6.626 \times 10^{-34} \times 2.998 \times 10^8 \times 6.022 \times 10^{23}}{1000 \times 498} \text{ m} = 2.40 \times 10^{-7} \text{ m}$$

(iii) $2.40 \times 10^{-7} = 2.40 \times 10^{-7} \times 10^9$ nm = 240 nm.
This is short-wavelength UV radiation.

Q4.5   1 kg of CFC-12 contains 1000/120.91 moles of CFC-12 = 8.2 moles. (The factor of 1000 is to convert kg to g.) If 1 molecule of CFC-12 destroys 1000 molecules of ozone, then 8.2 moles of CFC-12 will destroy 8200 moles of ozone. The molar mass of ozone is 48 g mol^{-1} and, therefore, the mass of ozone destroyed will be 8200×48 g = 394 $\times 10^5$ g = 394 kg.

Q4.6 Photochemical smogs are worst during the summer months because of the greater intensity of UV radiation as the sun is higher in the sky. Also, the days are longer in length. As Shanghai is in the northern hemisphere, June is in summer and December is in winter and, therefore, photochemical smogs will be worse during June than December.

Q4.7 (i)   A wind speed of 15 km hr^{-1} = 15 × 1000/3600 m s^{-1} = 4.2 m s^{-1}.

An emission rate of 5 kg hr^{-1} = 5 × 1000/3600 g s^{-1} = 1.4 g s^{-1}.

(ii)   Using Equation 4.35:

$$C = \frac{S \times l}{u \times h} = \frac{1.4 \times 1000}{4.2 \times 500} \text{ g m}^{-3} = 0.67 \text{ g m}^{-3}$$

(iii) The molar mass of SO_2 = 64.07 g mol^{-1}.
The number of moles of SO_2 in each m^3 = 0.67/64.07 mol m^{-3} = 0.0105 mol m^{-3}.

(iv) This number of moles will occupy 0.0105 × 22.4 L = 0.235 L.

(v)   The concentration of SO_2 therefore is 1000 × 0.235 ppmv = 235 ppmv.

Index

2,4,5,-T, 115

A

Acasta Gneiss, 74
Achondrites, 74
Acid dissociation constant, 129
Acid hydrolysis, 108
Acid mine drainage, 132, 155
Acid rain, 125, 132, 175, 184, 197
    effects of, 137
Acid-base titrations, 134, 161
Acids, 126, 161
    relative strength of, 127
Actinides, 9
Activation energy, 63
Aerosols, 165, 185
Agent Orange, 115
Agricultural pollutants, 159
Air pollution, 165, 184, 197
Airglow, 167
Albedo, 177
Albite, 83
Alcohols, 48
Aldehydes, 48
Alkali feldspar, 84
Alkali metals, 10
Alkaline earth metals, 10
Alkalinity, 144
Alkanes, 47
Alkenes, 48
Alkynes, 48
Alpha decay, 15
Aluminosilicates, 81, 98
    framework, 83
Aluminum, 73, 108
    structure of, 60
Amides, 49
Amines, 49
Ammonia, 122, 174
    anthropogenic sources of, 187
    structure of, 54
Ammonification, 174
Amoco Cadiz, 159
Amorphous solids, 25
Amount of substance, units of measurement for, 32

Amphibasic substances, 127
Amphiboles, 77
Amphiprotic substances, 127
Andalusite, 104
Andalusite-kyanite-sillimanite, phase diagram for, 105
Anhydrite, 60, 101
Anions, 41
Anomalous expansion of water, 121
Anorthite, 83
    phase diagrams for, 93
Anoxic environments, 150
Anthropogenic activities, 114, 197
Aqueous solutions
    analysis of metals in, 160
    units of concentration for, 33
Aragonite, structure of, 60
Argon, atmospheric levels of, 168
Argonite, 99
Aromatic compounds, structures of, 50
Arsenic, tube well contamination by, 157
Asteroid belt, 74
Atmosphere, 197
    chemical composition of, 167
    outgassed, 19
    oxygen in, 37
    primary, 168
    secondary, 168
    structure of, 165
Atomic mass, 5
Atomic number, 2
Atomic orbitals. *See also* orbitals
    overlap of, 40
Atomic spectral emission, 10
Atomic symbol, 2
Atomic weight, 5
    units of, 31
Atoms, 1, 67
Aurora Australis, 165
Aurora Borealis, 165
Auroras, formation of, 166

B

Bacteria, 110
Barringer Meteor Crater, 83, 96
Basalts, 73, 77

221